Bright Ideas for a Brighter Future: LEDs and the Fight Against Climate Change

Fader

Table of Contents

Chapter 1

Introduction

1.1 Motivation

Estimated by the United Nations Environmental Program, in 2010, lighting-purpose energy usage accounted for more than 15% of the global electricity consumption and almost 5% of the global green-house gas emission [1]. Energy consumption for illumination purposes is projected to increase monotonically, if the conventional lighting technologies continue to dominate the market [2]. Solid-state lighting (SSL) technology, in particular, light emitting diodes (LEDs) offers high efficiency, low cost, reliable light sources for not only general illumination but also applications, for example, in patient care and horticulture [3]. Rapid advancement and adoption of the SSL technology is projected to save up to 75% of the annual light-energy consumption as compared to a 'no-SSL' scenario by the year of 2035 [2].

An optimal white SSL source would require monolithic integration of blue, green, amber and red LEDs with high quantum efficiency. Till today, the efficiency of such color-mixed LEDs is still low due to the challenge of 'green gap', which refers to the low quantum efficiency (QE) of the LEDs emitting in the green and amber spectral region. As of 2017, external quantum efficiency (EQE) values of 80% and 44% were achieved in

InGaN quantum well (QW) based blue and green LEDs, respectively. On the other hand, EQE values of 63% and 18% have been achieved in AlInGaP based red and amber LEDs, respectively [2].

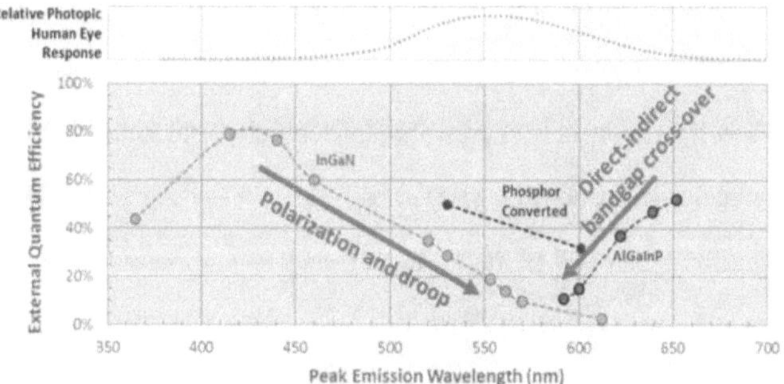

Figure 1.1 External quantum efficiency of state-of-the-art LEDs with different peak emission wavelength showing the 'green-gap.' This figure is adopted from Ref. [3].

As shown in Fig. 1.1, the QE of AlInGaP based LEDs drastically drops as the peak emission wavelength approaches amber and yellow from red, which is caused by direct-to-indirect bandgap cross-over in AlInGaP around 2 eV [4]. This fundamental limitation of group III-V based materials makes group-III-Ns a more viable candidate to be the active layer material in color-mixed white LEDs based on a single material platform. However, as shown in Fig. 1.1, the efficiency of InGaN QW based LEDs drops drastically as the peak emission wavelength extends from blue towards green, amber and red. One fundamental challenge in improving the radiative efficiency of these light emitters is related to the characteristic large polarization induced electric field of III-N semiconductors grown along the preferred c-plane orientation. Polarization induced electric field creates charge

2

separation resulting in a reduction in radiative recombination rate [5] of charge carriers. The detrimental impact from the internal electric field becomes more severe for InGaN LEDs emitting in wavelength beyond blue and green, in which higher-In content InGaN and relatively thicker QWs are required [6].

One approach to overcome the issue of polarization-induced charge separation in InGaN light emitters is to grow the structures along the non-polar (a- and m- planes) or semi-polar orientations [7, 8]. Separately, novel band-engineered QW structures such as staggered InGaN QW [9, 10, 11, 12, 13], strain-compensated InGaN-AlGaN QW [14, 15], type-II InGaN-GaAsN QW [16, 17], InGaN-delta-InN QW [18], and InGaN QW with delta-AlGaN layer [19, 20] were proposed to address the charge separation issue in conventional InGaN QW LEDs. Nevertheless, these QW designs still suffer from low efficiency when the emission wavelength extends to the longer wavelength regime, since higher In-content is still required in these QW designs. In another approach, rare-earth doped (Er [21], Eu [22]) GaN was studied as active region for green and red emission. However, the internal quantum efficiencies of these emitters are still below 1% [23].

1.2 ZnGeN$_2$ and ZnSnN$_2$ – two potential candidates for applications in high efficiency light emitters

The persistent challenges in improving the quantum efficiency of III-nitride LEDs for green and longer wavelengths call for the development of new materials. Among several candidates, two ternary heterovalent nitride compounds, namely, zinc germanium nitride (ZnGeN$_2$) and zinc tin nitride (ZnSnN$_2$) are especially promising thanks to their favorable band-alignment (large valence band offset) with GaN. The class of ternary heterovalent II-IV-N_2 compounds encompasses a range of nitride materials having two different elements in the cation sublattice – one from group-II (Be, Mg, Zn, Cd etc.) and the other from group-IV (Si, Ge and Sn). These materials are closely related to wurtzite III-N and have wurtzite derived crystal structures [24]. In an ideal II-IV-N_2 crystal, every nitrogen atom is tetrahedrally bonded with two group-II atoms and two group-IV atoms. Such an ordered arrangement of cations preserves the octet-rule. An ideal octet-rule-preserving unit cell of a II-IV-N_2 compound have orthorhombic symmetry of either space group $Pna2_1$ (sometimes referred to $Pbn2_1$, depending on the particular choice of principal axes), or $Pmc2_1$ [25]. Deviation from the ideal coordination of the cations around N will result in a locally octet-rule-violating wurtzite structure (space group $P6_3mc$) [25]. Although the octet-rule-violating, fully cation-disordered wurtzite structure has significantly higher predicted formation energy per formula unit than the ordered $Pna2_1$ or $Pmc2_1$ structures [26], this situation may be achieved, for example, in a kinetically limited growth regime [27]. The bandgap of ordered II-IV-N_2 materials span a range from ~0.64 eV ($CdSnN_2$) to 6.8 eV ($BeSiN_2$) [28]. However, the presence of disorder in the cation sublattice has been predicted to shrink the bandgap, so far in case of, $ZnGeN_2$, $ZnSnN_2$, and $MgSnN_2$, which can be a potential tool of bandgap tuning without alloying [26, 29, 30]. On the other hand, reduced crystalline symmetry of the ordered structure may induce non-linear optical properties. In addition, the presence of two elements of two different valencies could provide flexibility in doping and potentially enable achieving high p-doping level that is difficult in ternary III-N alloys [31].

1.2.1 Technologically important properties of $ZnGeN_2$ and $ZnSnN_2$

$ZnGeN_2$ and $ZnSnN_2$ have recently been attracting research interest for potential applications in optoelectronics [32, 33, 34, 35] and photovoltaics [25, 26, 36], thanks to their complementary properties to GaN [37]. For $ZnGeN_2$, experimentally determined bandgaps (3.36 – 3.4 eV) [38, 39] are in reasonable agreement with predicted bandgaps of the orthorhombic $Pna2_1$ ($Pbn2_1$) structure [37]. The experimentally determined bandgaps of $ZnSnN_2$ range from 1.7 eV-2.38 eV [40, 41, 42]. Some of these values are larger than the predicted bandgap (~1.8 eV for $Pna2_1$ structure [37]), which have been attributed to the Burstein-Moss shift in films with degenerate carrier concentration [41, 42]. The lattice mismatch of the $Pbn2_1$ $ZnGeN_2$ structure with GaN is less than 1%, whereas $ZnSnN_2$ is approximately lattice matched with $In_{0.31}Ga_{0.69}N$ [37]. One interesting aspect of these two materials is their band alignment with GaN. From the theoretical prediction based on first-principles calculations, the valence band and conduction band of $ZnGeN_2$ can be as high as 1.1 eV and 1 eV above those of GaN, respectively, at the heterointerface [43, 44]. For $ZnSnN_2$, these values are ~1.4 eV and -0.3 eV, respectively [43, 44]. The combination of closely lattice matching and large band offset of these Zn-IV-N_2 materials with III-Ns provides a unique opportunity for band structure engineering and to resolve the charge separation issue in nitride-based LEDs and thus, to potentially fill in the green-gap.

A $InGaN/ZnGeN_2$ heterostructure QW based design has already been studied for blue and green emission [32]. In this work, we investigated, for the first time, an $InGaN/ZnSnN_2$ heterostructure QW for amber light emission. The large valence band offset of $ZnGeN_2$ and $ZnSnN_2$ with GaN provides these novel QW structures with two essential advantages over conventional InGaN QW LEDs – (1) lowering In content requirement for achieving a desired peak emission wavelength and (2) strong hole confinement in the Zn-IV-N_2 layer yielding a remarkable enhancement of electron and hole wavefunction overlap, which in turn, increases the spontaneous emission rate. A lower In content requirement would allow higher growth temperature, and thus reduce the nonradiative recombination rate, which in turn further enhance the QE. The InGaN/ZnSnN$_2$ structure has one important advantage over the InGaN/ZnGeN$_2$ structure. In case of the latter, a large conduction band offset between ZnGeN$_2$ and InGaN pushes the electron wavefunction away from the ZnGeN$_2$ layer; this curbs the radiative recombination rate, which becomes more severe with increase in In content. On the other hand, ZnSnN$_2$ has very small conduction band offset with InGaN even for the In content that is required for the longest visible wavelength emission. This allows harnessing of the unmitigated benefits of the strong hole confinement provided by the large valence band offset. Therefore, a larger enhancement of electron-hole wavefunction overlap can be achievable in InGaN/ZnSnN$_2$ heterostructure QWs than in the one with ZnGeN$_2$, especially for amber and red emission. Nevertheless, both structures could provide improved radiative performance as compared to conventional InGaN QW LEDs over the entire visible spectral range.

1.2.2 Feasibility of implementation of III-N/Zn-IV-N$_2$ heterostructure devices

Although the efficacy of the novel InGaN/Zn-IV-N$_2$ (IV = Ge or Sn) heterostructure QWs for LEDs emitting beyond blue has been predicted numerically, realization of such structures is yet to be done. Metalorganic chemical vapor deposition (MOCVD) is the commercial growth technique of today's InGaN based LEDs. Therefore, it is technologically important to develop MOCVD growth for implementation of the novel InGaN/Zn-IV-N_2 based LEDs. For successful growth of a heterostructure device stack, growth conditions of different layers need to be compatible. For $ZnSnN_2$, the reported temperature ranges for $ZnSnN_2$ thin film deposition are 35 – 340 °C [45] for radio frequency (rf) sputtering, 350 – 550 °C by molecular beam epitaxy (MBE) [46, 47]. Thermal decomposition of ammonia (NH_3), the commonly used MOCVD precursor for N, starts at ~450 °C imposing a lower bound for growth temperature of nitride materials in conventional MOCVD system [48]. In addition, a lower growth temperature usually results in a larger density of defects in InGaN QW. Therefore, MOCVD growth of InGaN/$ZnSnN_2$ heterostructure QW using thermally decomposed NH_3 can be challenging. A probable solution can be phonon-assisted decomposition of NH_3, which has been demonstrated to be useful for low temperature MOCVD growth of GaN [49].

ZnGeN$_2$ has been reported to be stable up to 950 °C and has been reported to be synthesized at a wide range of growth temperature (500 °C - 1000 °C). On the other hand, commonly used growth temperature for InGaN is between 650°C - 750 °C. Therefore, from implementation point of view, an InGaN/$ZnGeN_2$ heterostructure is more feasible than an InGaN/$ZnSnN_2$ heterostructure using the conventional MOCVD growth method.

1.2.3 Synthesis and characterization of ZnGeN$_2$

To date, there has been only a handful of reports on the synthesis of $ZnGeN_2$ by different growth techniques including chemical vapor deposition (CVD) [50], halide vapor phase epitaxy (HVPE) [51], high pressure synthesis [52], radio frequency sputtering [53], vapor-liquid-solid [27, 39, 54], ammonothermal [55], MOCVD [56, 57] and MBE [58]. The reported growth temperatures range from 500 to 1000 °C. The ordering of cations was achieved by higher growth temperatures [52] or post growth annealing [27]. Experimental investigations of $ZnGeN_2$ have mainly focused on the morphological [57, 58, 59, 60, 61], optical [38, 39, 59, 60], crystal structural [38, 56, 57, 58, 59, 60, 61], and lattice vibrational properties [27, 54, 62]. Deep-level defects in MOCVD-grown $ZnGeN_2$ films on sapphire substrates have also been investigated [63]. Recently, valence band offset of $ZnGeN_2$ and $(ZnGe)_{0.94}Ga_{0.12}N_2$ with GaN were determined experimentally by our group [64].

1.4 Organization of the book

In this book, $InGaN/Zn$-IV-N_2 heterostructure QWs have been investigated theoretically and experimentally.

Chapter 2 of this book discusses about general properties of ternary heterovalent II-IV-N_2 compounds. The crystalline structures and lattice parameters, electronic band structures and bandgaps, band alignment with GaN, cation disorder and their effects, native defects and potential dopants, and lattice vibrational properties have been discussed.

A design of an $InGaN$-$ZnSnN_2$ based QW for high efficiency amber LEDs is presented in Chapter 3 of this book. Chapter 4-7 of this book presents the MOCVD growth of $ZnGeN_2$ and $InGaN/ZnGeN_2$ heterostructure QWs. Development of MOCVD growth window for stoichiometric $ZnGeN_2$ films is a prerequisite for the successful implementation of the heterostructure LED. We have conducted a systematic investigation of the MOCVD growth of $ZnGeN_2$ films grown on GaN templates and sapphire substrates. The work on MOCVD growth of $ZnGeN_2$ films on differently oriented sapphire substrates is presented in Chapter 4. By varying the growth conditions, single crystalline $ZnGeN_2$ films were obtained on c-, r-, and a-sapphire substrates. The crystalline, morphological, optical and electrical transport properties of the grown films were investigated. In addition, thermal annealing studies have been conducted on $ZnGeN_2$ films grown on c-sapphire substrates.

Key growth parameters such as growth temperature, total reactor pressure, and cationic precursor flow rates were mapped to determine their impacts on the cation stoichiometry of $ZnGeN_2$ films grown on GaN/c-sapphire templates. Extensive material characterization was performed to investigate the structural, morphological, optical and transport properties of the films. The Zn/(Zn+Ge) compositions in the films have been found to depend on combination growth temperatures, reactor pressure and precursor molar flow rates. The surface morphology and the crystalline properties of $ZnGeN_2$ films showed dependency on the cation stoichiometry. The results are presented in Chapter 5.

The valence band offset of $ZnGeN_2$ and $(ZnGe)_{0.94}Ga_{0.12}N_2$ with GaN was determined, for the first time, which is presented in Chapter 6 of this book. Implementation of $InGaN/ZnGeN_2$ heterostructure QW structures are discussed in Chapter 7. The structures of the grown QWs as well their emission properties have been investigated. Finally, Chapter 8 concludes this book and lays out an outlook of the future research directions on this topic.

Chapter 2

Ternary heterovalent II-IV-N_2 compounds

Ternary heterovalent II-IV-N_2 compounds are closely related to group-III-N compounds. Hypothetically, the ideal II-IV-N_2 structure can be derived from a III-N crystal by replacing every two group-III atoms by one atom of a group-II element and one atom of a group-IV element. These ternary heterovalent compounds have similarities with their binary III-N counterparts. For example, the crystal structures of ternary II-IV-N_2 compounds are closely related to the binary III-N. However, the presence of two different cations in II-IV-N_2 causes significant deviations in some properties from the respective binary III-Ns. A comparative discussion on fundamental properties of II-IV-N_2 vs III-N compounds is presented in this chapter.

2.1 Crystal structures and lattice parameters

For group III-N semiconductors, the most stable phase has a wurtzite structure (space group $P6_3mc$). Symmetry is absent in a wurtzite lattice along c-direction, which results in strong polarization in these materials. The III-N derived II-IV-N_2 has an orthorhombic symmetry of $Pna2_1$ space group with $a \approx \sqrt{3}a_w$, $b = 2a_w$ and $c = c_w$, where a_w and c_w refer to the wurtzite lattice parameters. Therefore, the $Pna2_1$ structure is a $\sqrt{3} \times 2$ superlattice along the ortho-hexagonal axes of the $P6_3mc$ structure [24, 65]. Interchanging

10

the a and b axes of the Pna2₁ structure would result in an equivalent Pbn2₁ structure. An equivalent unit cell with a = 2a$_w$, b ≈ √3a$_w$ and c = c$_w$. Figure 2.1, reproduced from Ref. [65], shows relationship between the Pbn2₁ orthorhombic and wurtzite crystal structures. Pbn2₁ structure is more commonly used in first principles-based calculations [24, 37]. Both structures have sixteen atoms in the unit cell.

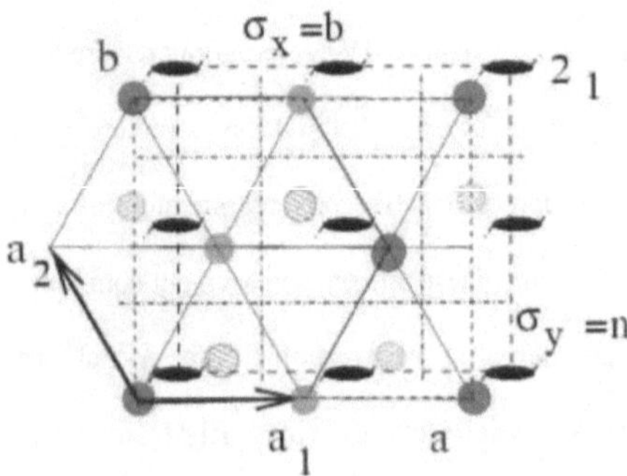

Figure 2.1 Relationship between Pbn2₁ orthorhombic and P6₃mc wurtzite structures. This figure is reproduced from Ref. [37].

In an orthorhombic II-IV-N₂ structure, the tetrahedral bond length of group-II atom with N are smaller than that of group-IV atom with N. This leaves two inequivalent positions of N atoms in the unit cell [24]. Based on first principles calculations for Zn-IV-N₂ compounds, the Zn-N bond length remains almost constant (~2.02 - 2.06 Å) whereas the IV-N bond length increases with increase in the atomic number of the group-IV element (1.76 Å in ZnSiN₂, 1.88 Å in ZnGeN₂, and 2.02 Å in ZnSnN₂) [65]. Similar trend has been reported for Mg-IV-N₂ [66] and Cd-IV-N₂ [67] compounds. The 2b/a parameter for the Pbn2₁ (2a/b for Pna2₁) is an indication of how close the structure is to the ideal wurtzite structure. For an ideal wurtzite structure, the 2b/a value would be √3 (≈1.732). In case of the Zn-IV-N$_2$, the experimentally reported values of 2b/a are 1.731 for ZnSiN$_2$, 1.693 for ZnGeN$_2$ and 1.73 for ZnSnN$_2$ [40]. For MgSiN$_2$ and MgGeN$_2$, the values are 1.628 and 1.661, respectively; experimental value for MgSnN$_2$ has not been reported in literature. Corresponding predicted values by first principles calculations are in good agreement with the experimental values [66, 68].

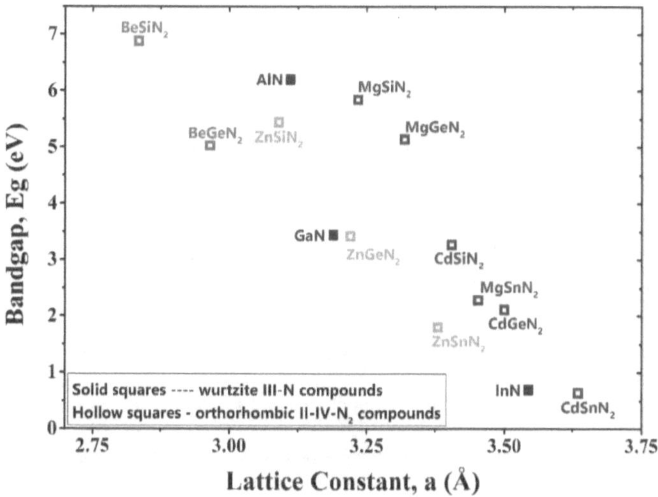

Figure 2.2 Bandgap vs wurtzite lattice constant plot of III-N and II-IV-N$_2$ [28] compounds. Please note that there are significant uncertainties in the bandgap and lattice constant values of II-IV-N$_2$ in literature.

The wurtzite lattice parameter of orthorhombic $MgSiN_2$ and $ZnGeN_2$ are very close to that of GaN. In addition, the wurtzite lattice parameters of $MgGeN_2$, $CdSiN_2$, $ZnSnN_2$, $MgSnN_2$ and $CdGeN_2$ lie between those of GaN and InN meaning they are lattice-matched with $In_xGa_{1-x}N$ for appropriate In content [28]. Figure 2.2 shows the bandgap vs. wurtzite lattice constants of III-N and II-IV-N_2 compounds.

Besides $Pna2_1$ (or $Pbn2_1$), another structure with eight atoms per unit cell and $Pmc2_1$ symmetry has been predicted to be stable by first principles calculations [25, 26, 36]. Energy of formation of this smaller unit cell structure is very close to the $Pna2_1$ structure in many cases [36].

2.2 Electronic band structures and bandgaps

Based on first principles calculations, the orthorhombic II-IV-N_2 compounds cover the bandgap ranging from 0.64 eV ($CdSnN_2$) to 6.8 eV ($BeSiN_2$) [28]. This range is slightly larger than that of the wurtzite III-Ns. The top valence band of II-IV-N_2 compounds are more of N-2p like, whereas the lowest conduction band is more of cation-s type [65]. In most of the II-IV-N_2s, the conduction band minimum has slightly higher group-IV s-character than group-II s-character. Consequently, a change in the group-IV element has larger effects on the bandgap as compared to a change in group-II element [37]. $MgSinN_2$, $ZnGeN_2$ and $CdSnN_2$ are the closest II-IV-N_2 analogs of AlN, GaN and InN, respectively, since both cations in these II-IV-N_2 compounds belong to the same row in the periodic table as those of respective group-III elements. The lowest calculated direct band gaps of these II-IV-N_2s (6.3 eV [66], 3.4 eV [37] and 0.64 eV [69], respectively) are very close to their III-N analogs. However, $MgSiN_2$ has been predicted to have a lower indirect bandgap of 5.85 eV [66], which is in good agreement with the measured value of 5.6±0.2 eV by X-ray emission and absorption spectroscopy [70].

Zn-IV-N_2s represent the most investigated II-IV-N_2 compounds. For $ZnGeN_2$, the measured absorption edge was found to be in the range of 2.67 eV to 3.3 eV [38, 51, 53], whereas near band-edge photoluminescence peaks were observed around 3.3 eV in MOCVD grown film on sapphire [38] and at 3.4 eV from a single crystalline rod [39]. The lower values of the absorption edge might have been caused by the defects in the films [24]. For $ZnSnN_2$, the measured absorption edge values vary even more from 1.7 eV from polycrystalline samples grown by vapor-liquid-solid technique [40] to 2.38 eV in single crystalline films grown by molecular beam epitaxy [42]whereas the calculated values lie in the range of 1.4 – 1.8 eV [41]. The higher absorption edge values have been attributed to the Moss-Burstein shift in band-edges due to the degenerate carrier concentrations in the films [41]. This scenario is similar to the earlier stages of research on InN. Moreover, the presence of octet-rule violating disorder in the cation sublattice has been found to significantly affect the bandgap of II-IV-N_2 materials results in further ambiguity in the bandgap of this material system.

The bandgap of $Pmc2_1$ structures are often very close to those of $Pna2_1$ structures [25, 29, 36].

2.3 Disorder in cation sublattice and its effects on optical properties

For III-Ns, disordered structures of isoelectric alloys are thermodynamically more favorable than the ordered structures like the alloys of binary III-Vs. For InGaN and AlGaN alloys, chemical ordering can be achieved under kinetically limited surface driven growth conditions [71, 72] for example, high growth temperatures or low growth rates can enable the adatoms to diffuse to the preferred incorporation sites [72]. Chemical ordering of the cations in AlGaN or InGaN reduces the bandgap.

In clear contrast to the AlGaN or InGaN alloys, ternary heterovalent II-IV-N_2s are most stable in the cation ordered orthorhombic $Pna2_1$ structure. However, as mentioned earlier, another ordered structure but with a smaller unit cell and $Pmc2_1$ symmetry has been predicted to have very close formation energy per unit cell [25, 41]. Both these are locally octet rule preserving structures in which each N atom is coordinated by two group-II and two group-IV atoms; in fact, they are just structures with different stacking of group-II and two group-IV cations in the basal plane [26]. Deviation from ideal (2,2) coordination of N atoms by the cations would result in an octet-rule violating disordered structure. An octet rule preserving disordered structure can also be achieved by simply random stacking of $Pna2_1$ and $Pmc2_1$ structures [25]. Both disordered structures have wurtzite symmetry.

As mentioned earlier, in an ideal wurtzite like structure of II-IV-N_2s, the 2a/b ratio (for $Pnn2_1$ configurations of principal axes) is 1.73. A deviation occurs in the orthorhombic structure due to different bond lengths of group-II and group-IV atoms with N. Therefore, like the c/a ratio in case of the II-IV-V_2s, the 2a/b ratio is a predictor of the likelihood of a II-IV-N_2 compound to have cation disorder. For example, $ZnSnN_2$ (2a/b = 1.730) is more likely to have cation disorder than $ZnGeN_2$ (1.693), which is consistent with the experimental results on these two materials. Table 2.1 lists the 2a/b ratio for different II-IV-N_2 compounds.

Table 2.1 b/a_w for Mg-IV-N$_2$, Zn-IV-N$_2$ and Cd-IV-N$_2$ compounds. The asterisk sign indicates the experimental values [65, 66, 69].

Material	b/a_w
MgSiN$_2$	1.628*
MgGeN$_2$	1.661*
MgSnN$_2$	1.718
ZnSiN$_2$	1.731*
ZnGeN$_2$	1.693*
ZnSnN$_2$	1.730*
CdSiN$_2$	1.600
CdGeN$_2$	1.630
CdSnN$_2$	1.690

Since the bandgap of Pna2$_1$ and Pmc2$_1$ structures are similar, the octet rule preserving disordered structure resulting from their random ordering would have a similar bandgap also. However, octet-rule violating disorders can substantially shrink the bandgap, for example, up to almost 3 eV in ZnGeN$_2$ or can even close the bandgap, for example in ZnSnN$_2$ [25, 26]. The reduction in the bandgap has been attributed to an intermediate band close to the valence band but inside the gap which can be caused by the large concentration (e.g., $\sim 10^{20}$-10^{21} cm^{-3} in ZnGeN$_2$) of the exchange defects in the crystals. Exchange defects refer to the swap of group-II and group-IV atoms in the sublattice, which results in (3,1) or (1,3) coordination. In the case of complete disorder, one would expect binomial distribution N atoms having (4,0), (3,1), (2,2), (1,3) and (0,4) coordinations. However, (4,0) and (0,4) coordinations are highly unlikely because of their high formation energy. Such high density of defects can be formed during growth under highly non-equilibrium conditions such as in kinetically limited growth regime [27].

2.4 Native defects and impurity doping

For III-Ns, Si and Mg have been the most commonly used n- and p-type dopant. However, the level of achievable hole concentrations has been limited by the activation energy of Mg, which is close to 140 meV.

There has been no report on extrinsic doping of II-IV-N_2 compounds. However, unintentional n-type conductivity in $ZnGeN_2$ (electron concentration $\sim 10^{18}$-10^{19} cm^{-3}) [51, 59] and $ZnSnN_2$ (electron concentrations $\sim 10^{18}$ -10^{21} cm^{-3}) [45, 41] have been reported. For $ZnSnN_2$, post-growth annealing was found to be effective in reducing the electron concentrations by almost one order of magnitude [73]. Carrier mobilities up to 10 cm^2/V·s and 17 cm^2/V·s were obtained in $ZnSnN_2$ film grown on yttria stabilized zirconia [42] and $ZnGeN_2$ films grown on sapphire [59], respectively. These mobility values are limited by the poor crystalline quality caused by the large lattice mismatch between the film and the substrate. The effective mass of electrons and holes in II-IV-N_2 compounds are of the same order as those of GaN [37, 66, 69]. However, there has not been any report on theoretical limits of the carrier mobility of II-IV-N_2s to the best of our knowledge. Growth on lattice matched substrates can potentially reduce the structural defects and thus, improve the carrier mobility.

Based on first principles calculations by Skachkov et al. [74], the cation antisite defects, namely, Ge-at-Zn (Ge_{Zn}) and Zn-at-Ge (Zn_{Ge}) are the dominant native defects in

$ZnGeN_2$, which are of donor and acceptor type, respectively. $ZnGeN_2$ grown under equilibrium conditions has been predicted to be semi-insulating [28]. This is an important difference with GaN, which shows n-type conductivity in the absence of intentional doping. Substitutional O at the anion site (O_N) is another potential unintentional donor in $ZnGeN_2$ or $ZnSnN_2$ [45, 74]. The formation of the native defects can be affected by the chemical potential of the growth species [74, 75].

There have been few theoretical investigations on potential n- and p-type dopant in Zn-IV-N_2 compounds [31, 76, 77]. For $ZnSiN_2$ and $ZnGeN_2$, P can be a potential shallow donor when it substitutes the group-IV atoms (Si and Ge). However, compensation by the native acceptors can prevent achievement of successful n-type doping in $ZnSiN_2$. For $ZnGeN_2$, P_{Ge} is predicted to provide electron concentrations up to 10^{19} cm^{-3}. S and Se at the anion site and As at the Ge site can also act as the shallow donor in $ZnGeN_2$. One important aspect of few of these n-type dopants, as compared to the scenario of Si doping in AlGaN, can be the absence of D-X behavior. D-X behavior refers to the tendency of any shallow donor impurity to undergoing a large lattice relaxation and converting to a deep accepter by capturing two electrons [31]. In high Al composition AlGaN, Si has a transition level between +1 and -1 charge states that lies inside the bandgap and can possibly undergo the transition and act as acceptors, which limits the achievable doping level in this alloy [31]. Based on first principles calculation, P_{Ge}, As_{Ge}, S_N and Se_N in $ZnGeN_2$ and P_{Si} in $ZnSiN_2$ did not show D-X behavior [31].

Regarding the p-type doping, first principles calculations indicates that Li substitutional impurity at Zn site (Li_{Zn}) can be a shallow acceptor in $ZnSnN_2$ [78]. On the other hand, there have been two contradicting theoretical reports regarding Li doping in ZnGeN$_2$ - one suggesting Li$_{Zn}$ to be a shallow acceptor [76] and the other to be a deep donor [77]. Nevertheless, in Ref. [76], the activation energy of Al$_{Ge}$ has been found to be 240 meV, representing the minimum among the potential candidates and comparable to the activation energy of Mg acceptor in GaN. However, group-III atoms can also take the Zn site, especially in Zn poor growth conditions, which will be a donor type substitution and is anticipated to compensate the p-type doping. It has been suggested that co-doping with H during the growth can prevent the Al$_{Zn}$ substitution; H can be removed by post-growth annealing and thus successful p-type doping can possibly be achieved [76].

Figure 2.3 The natural band alignment of orthorhombic II-IV-N$_2$s with wurtzite III-Ns and ZnO determined by Lyu et al. [79] using first principles surface calculations of their electron affinity. This figure is reproduced from Ref. [79].

2.5 Band-alignment

For Mg-IV-N$_2$, Zn-IV-N$_2$ and Cd-IV-N$_2$ compounds, the bandgap decreases with increase in the atomic number of the group-IV cation, while the band alignments remain type-I [79]. Type-I band alignments retain in II-Si-N_2 and II-Ge-N_2 compounds as well; however, MgSnN$_2$ has a type-II staggered band alignment with ZnSnN$_2$. Of technological interest is the band alignment of ZnSnN$_2$ and ZnGeN$_2$ with GaN [32, 34, 35]. For both these Zn-IV-N_2 compounds, the valence band is more than 1 eV above than that of GaN at the heterointerface [43, 44]. Figure 2.3 (reproduced from Ref. [79]) depicts the band alignments of different II-IV-N_2 compounds with III-Ns.

Band alignment of ZnGeN$_2$ relative to GaN is anomalous with those of ZnGeP$_2$, ZnGeAs$_2$ and ZnGeSb$_2$ with GaP, GaAs and GaSb, respectively. The ZnGe-V_2 compounds have type-I band alignment with respective analogous binary Ga-V (V = P, As, Sb) compounds whereas band alignment of ZnGeN$_2$ with GaN is of type-II. This has been attributed to the small cation-N band length as compared to cation-V (V = P, As, Sb) bond lengths [80].

2.6 Vibrational modes

The presence of two cations causes significant deviation in the lattice vibrational properties of II-IV-V_2 compounds as compared to their III-V counterparts [81]. As compared to only two optical modes in binary zincblende III-Vs, a chalcopyrite II-IV-V_2 has fifteen vibrational modes. Of these four are Raman active, nine are both Raman and infrared active and two are inactive. Another significant difference is the polarization sensitivity of the vibrational modes in II-IV-V_2 compounds unlike the III-Vs [81].

In case of the nitrides, the vibrational properties of the II-IV-N_2 compounds are substantially different. Mg-IV-N_2 and Zn-IV-N_2 compounds have been reported to have total seventy-eight vibrational modes all of which are Raman active. Some of these modes can be related to those of the wurtzite structure by folding of Brillouin zone. However, the difference in the II-N and IV-N bond-lengths [24] needs to be considered. Moreover, polarization sensitivity of the vibrational modes of orthorhombic II-IV-N_2 structures has been reported both using first principles calculations as well as experimentally [54, 62, 68]. The differences in vibrational mode have been predicted to affect the thermodynamic properties, for example, heat capacity and thermal conductivity [82].

In a number of experimental studies on $ZnGeN_2$ and $ZnSnN_2$, only phonon density of states like features were observed in the Raman spectra, which have been attributed to the structural defects or the presence of cation disorder [54, 62]. Probing the vibrational modes can be a potential technique for identifying the cation disorder in II-IV-N_2s.

2.7 Polarization properties and non-linear optical properties

The absence of inversion symmetry causes large polarization in wurtzite crystals along c-axis. The spontaneous polarization in Zn-IV-N_2 compounds were found to be comparable to that of GaN [68]. Interestingly, the differences between spontaneous polarization of Zn-IV-N_2 compounds are much smaller than those between III-Ns. The piezoelectric coefficients were found to be stronger in $ZnSiN_2$ and $ZnSnN_2$ than in $ZnGeN_2$, which have been attributed to the stronger ionicity of the bonds in these materials. The reduced symmetry in chalcopyrite II-IV-V_2 compounds results in their non-linear optical properties. The number of studies on non-linear optical properties on II-IV-N_2s have been limited. One study on $ZnGeN_2$ suggested that this material is probably not suitable for non-linear optics application [24].

2.8 Conclusions

In summary, the II-IV-N_2 compounds have a wurtzite derived crystal structure, which has the orthorhombic symmetry in the most stable phase. The bandgap range of II-IV-N_2 compounds is slightly larger than the bandgap range of III-Ns. For the II-IV-N_2, the cation ordered phase is the most stable one unlike the III-Ns. The presence of octet-rule violating cation disorder significantly shrink the bandgap of II-IV-N_2 unlike the III-Ns in which the disordered phase has the highest bandgap. The lattice vibrational properties of II-IV-N_2s significantly differ from those of III-Ns. The complementary properties of II-IV-N_2 with GaN can be exploited for potential device applications.

Chapter 3

Design of InGaN/ZnSnN$_2$ based high efficiency LEDs beyond green

3.1 Introduction

In the past decades, the performance of InGaN based blue and green light emitters have been improved tremendously, although the efficiency of green LEDs is still inferior to that of the blue ones. In general, the efficiency of InGaN LEDs decreases as the emission wavelength extends from blue to green, amber and red. In particular, the efficiencies of amber LEDs have been the lowest among the visible LEDs to date [83]. One fundamental challenge in improving the radiative efficiency of InGaN QWs based light emitters originates from the characteristic large polarization induced electric field of III-nitride semiconductors grown along the preferred c-plane orientation. This leads to charge separation and resultant reduction in charge carrier radiative recombination rate [5]. The detrimental impact from the internal electrostatic field becomes more severe for InGaN LEDs emitting in wavelength beyond blue and green, in which higher-In content InGaN and relatively thicker QWs are required [6].

Approaches have been used to overcome the issue of polarization-induced charge separation in InGaN light emitter devices by growing the structures along the non-polar (a- and m- planes) or semi-polar orientations [7, 8, 83]. Separately, novel QW structures such

23

as staggered InGaN QW [9, 10, 11, 12, 13], strain-compensated InGaN-AlGaN QW [14, 15], type-II InGaN-GaAsN QW [16, 17], InGaN-delta-InN QW [18], and InGaN QW with delta-AlGaN layer [19, 20] have been proposed to address the charge separation issue in conventional InGaN QW LEDs. These novel QW designs using nanostructure engineering have shown improvements in radiative efficiency. However, these QW designs still suffer from low efficiency when the emission wavelength extends to the longer wavelength regime, since higher In-content is still required in these QW designs. In addition, rare-earth doped (Er [21], Eu [22]) GaN was studied as active region for green and red emission. However, the internal quantum efficiencies of these emitters have not crossed 1% yet [23].

Recently, an InGaN-ZnGeN$_2$ based type-II QW structure has shown great improvement in the radiative efficiency in blue and green LEDs [32]. GaN and ZnGeN$_2$ have similar bandgaps with a large band offset ($\Delta E_V \sim 1.1$ eV), which results in a type-II heterointerface between InGaN and ZnGeN$_2$ [24, 43, 44]. The large band offset in the valence band leads to a strong confinement of the hole wavefunction in the ZnGeN$_2$ layer. However, the similar large band offset in the conduction band pushes the electron wavefunction away from the ZnGeN$_2$ layer, which limits the radiative recombination rate between electrons and holes.

On the other hand, ZnSnN$_2$ represents the smallest bandgap ($E_g \sim 1.8$ eV) semiconductor among the Zn-IV-N$_2$. From the recent first principles calculations, ZnSnN$_2$ has a favorable alignment in both valence band ($\Delta E_V = 1.4$ eV) and conduction band ($\Delta E_C = -0.3$ eV) with GaN [43, 44]. This allows a strong confinement of hole wavefunction in the valence band, without sacrificing the shift of the electron wavefunction in the conduction band. Therefore, a large electron-hole wavefunction overlap for high internal quantum efficiency is achievable for LEDs with longer emission wavelength beyond green.

In this chapter, we investigate the design of an InGaN-ZnSnN$_2$ based QW structure for high efficiency amber LEDs using the self-consistent 6-band $k\cdot p$ method [14]. The polarization field, strain effect, and carrier screening effect were taken into consideration in the band structure calculation. The spontaneous emission radiative recombination rates are calculated for both the InGaN-ZnSnN$_2$ QW and the conventional InGaN QW emitting at the similar wavelength, which indicates a 270-320 X enhancement at different carrier concentration.

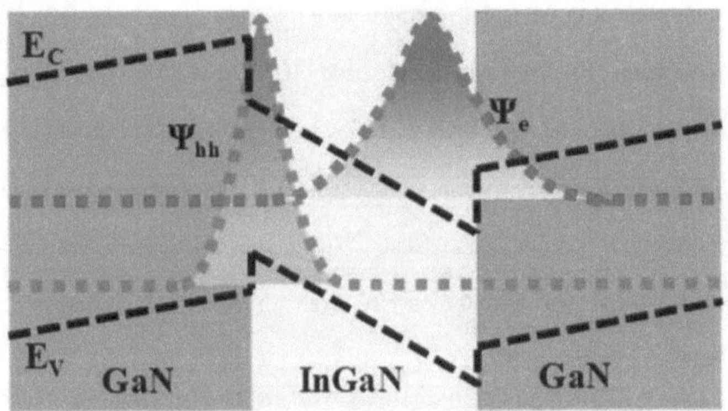

Figure 3.1 Illustration of charge carrier separation in conventional InGaN QW with GaN barriers due to large band-bending caused by polarization field. [35]

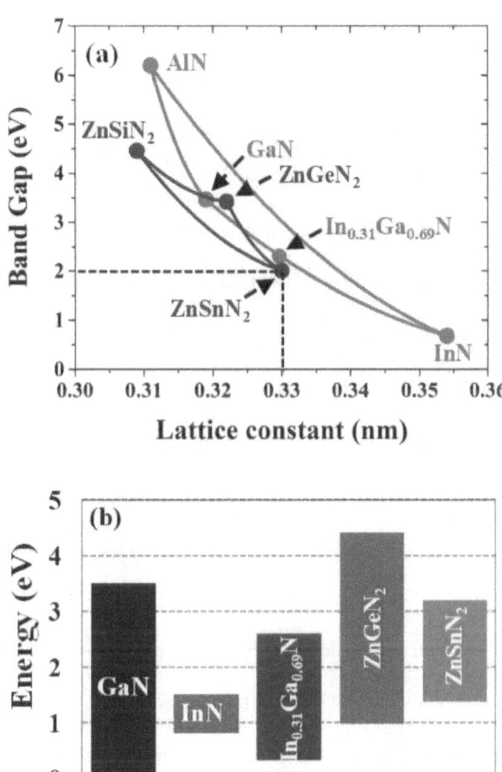

Figure 3.2 (a) Energy bandgap versus wurtzite lattice constant of III-N and Zn-IV-N$_2$ semiconductors. (b) Bandgap alignments of InN, In$_{0.31}$Ga$_{0.69}$N, ZnGeN$_2$ and ZnSnN$_2$ with respect to GaN showing conduction and valence band offsets. The values for Zn-IV-N$_2$ materials were obtained from Ref. [24]. This figure was published in Ref. [35]

3.2 Conceptual design

The absence of inversion symmetry in wurtzite III-nitride materials causes large spontaneous polarization whereas the lattice-mismatch induced strain in III-nitride heterostructures gives rise to piezoelectric polarization [6]. In InGaN based QWs with GaN as barriers, the polarization induced electric field creates band bending. As a result, charge carriers of opposite polarity are localized in separate spatial regions in the QW [84, 85, 86, 87, 88] and thus, decreases the electron-hole wavefunction overlap $\Gamma_{e\text{-}h}$, as shown in Fig. 3.1. The detrimental effect becomes more severe when higher In content InGaN or thicker QW are used targeting for longer wavelength emission.

In the proposed design, a thin layer of $ZnSnN_2$ is used as the hole confinement layer in the active region to address the charge separation issue. The energy bandgap versus wurtzite lattice parameter diagram of $Zn\text{-}IV\text{-}N_2$ and III-N as plotted in Fig. 3.2(a) reveals that $ZnSnN_2$ is lattice-matched to $In_{0.31}Ga_{0.69}N$ [24]. Most importantly, the band alignment of $ZnSnN_2$ with unstrained GaN and InN shown in Fig. 3.2(b) reveals that there is a large ΔE_V and a close-to-zero ΔE_C between $ZnSnN_2$ and InGaN [43, 44]. Such a band alignment of $ZnSnN_2$ with InGaN renders a strong hole confinement in the $ZnSnN_2$ layer but still maintains the uniform distribution of electron wavefunction in the conduction band. Note that, first principles calculations [68] suggest that $ZnSnN_2$ has a spontaneous polarization (see Table 3.1) that is comparable to that of GaN, which has been taken into account in this calculation. This QW design also allows to extend the transition wavelength into longer wavelength regime without using high-In InGaN.

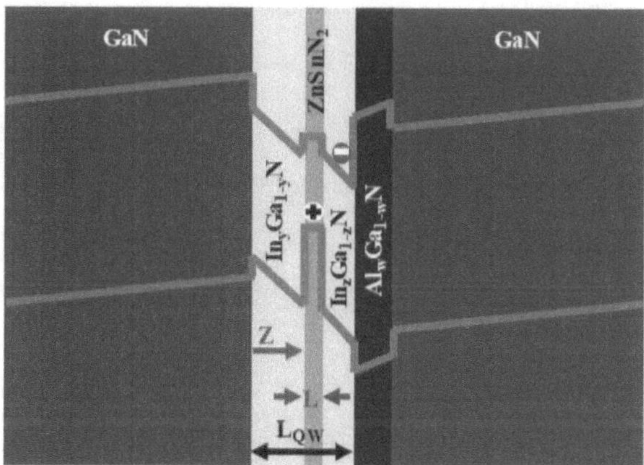

Figure 3.3 Schematic of the GaN/In$_y$Ga$_{1-y}$N/ZnSnN$_2$/In$_z$Ga$_{1-z}$N/Al$_w$Ga$_{1-w}$N/GaN QW showing the band edge alignment. + and − signs denote low energy regions for holes and electrons, respectively. Position of the conduction band edge of ZnSnN$_2$ with respect to that of InGaN depends on the In content. [35]

Figure 3.3 schematizes the conceptual design of an InGaN-ZnSnN$_2$ based QW structure with GaN and GaN/AlGaN as barriers. The purpose of the thin AlGaN layer is to better confine the electron wavefunction in the QW active region. While the peak emission wavelength $\lambda_{peak,}$ of a conventional InGaN QW depends on the In-content and the thickness of the QW, the InGaN-ZnSnN$_2$ based QW structure allows a greater flexibility and wider range of tuning for λ_{peak}. Particularly, the thickness and position of the ZnSnN$_2$ layer within the InGaN QW affect the confined hole energy levels and hence the λ_{peak}. Hereafter, the conventional GaN/In$_x$Ga$_{1-x}$N/GaN QW as shown in Fig. 3.1 and the proposed GaN/In$_y$Ga$_{1-}$

$_y$N/ZnSnN$_2$/In$_z$Ga$_{1-z}$/Al$_w$Ga$_{1-w}$N/GaN QW as shown in Fig. 3.3 will be referred to as IGN QW and IGN-ZTN QW, respectively.

The QW thickness and In content of the IGN QW were designed for peak emission wavelength at amber (600 nm). The total QW thickness L_{QW} of IGN-ZTN QW was kept the same as that of the IGN QW and the In content in the two InGaN sub-layers was kept the same.

3.3 Numerical formulation

A self-consistent 6-band k.p method was used to obtain the band structure of the proposed as well as the conventional QW structures. The hole energy bands were calculated using 6 x 6 diagonalized k.p Hamiltonian whereas the parabolic energy bands were assumed for electrons. Effects of strain, spontaneous as well as piezoelectric polarization, carrier screening effect and valence band mixing were taken into account. In this study, we mainly focus on QW structures emitting in the visible wavelength regime. This allows us to assume that the coupling between the conduction and valence bands are week. Therefore, the band structure in the conduction band is assumed as parabolic in the vicinity of the conduction band minimal. Many body effects and inhomogeneous broadening of In-content in the QW were not considered. Previous studies have shown that the many body effects, which include bandgap renormalization and the excitonic or Coulombic enhancement can affect the peak gain and corresponding wavelength [89]. Thus, the many body effect can affect the absolute value of the calculated spontaneous emission rate and the emission wavelength. However, this will not change the trend of the results, neither on the design.

Table 3.1 Material parameters of GaN, InN and ZnSnN$_2$ used in the simulation. The values were obtained from Ref. [90, 91] (GaN and InN) and Ref. [24, 43, 44, 86], and [65] (ZnSnN$_2$).

Parameter	GaN	InN	ZnSnN$_2$
Lattice constant (Å)			
a	3.189	3.545	3.3
c	5.185	5.703	5.462
Energy Parameters (eV)			
E_g at 300 K	3.42	0.6405	1.8
Δ_1 (=Δ_{cr})	0.01	0.024	0.088
$\Delta_1 = \Delta_2 = \Delta_{so}/3$	0.00567	0.00167	0
Conduction band offset with GaN (eV)		0.7 ΔE_g	-0.3
Valence band offset with GaN (eV)		0.3 ΔE_g	1.4
Conduction-band effective masse			
$m_{\parallel}^*/m_0$ at 300 K	0.21	0.07	0.13
$m_{\perp}^*/m_0$ at 300 K	0.2	0.07	0.17
Valence band effective mass parameters			
A_1	-7.21	-8.21	- 8.23
A_2	-0.44	-0.68	- 0.49
A_3	6.68	7.57	7.77
A_4	-3.46	-5.23	- 2.8
A_5	-3.4	-5.11	- 2.8
A_6	-4.9	-5.96	-3.89
Elastic stiffness coefficients (GPa)			
C_{11}	390	223	272
C_{12}	145	115	128
C_{13}	106	92	100
C_{33}	398	224	306
Spontaneous Polarization (C/m^2)	-0.034	-0.042	-0.029
Piezoelectric coefficients (pmV^{-1})			
d_{13}	-1	-3.5	-2.9
d_{33}	1.9	7	5.4

protocol of InGaN-ZnSnN$_2$ QWs for high efficiency amber LEDs. Inhomogeneous broadening of QW thickness or In composition in InGaN based light emitters have been reported to result in spectral broadening and shift as well as reduction in laser gain [92]. These effects become more obvious when high In content is used or higher carrier concentration needs to be considered in the laser operation. In this work, we have investigated the recombination properties at the relatively low carrier densities (1-5×10^{18} cm^{-3}) for LEDs application. In order to take into account these effects accurately, parameters from experiments are necessary, which is out of the scope of this study. However, this does not affect the effectiveness of using the concept of InGaN-ZnSnN$_2$ QWs for high efficiency amber LEDs.

The energy band alignments in the heterostructures were calculated by iteratively solving Poisson's equation until the convergence was reached. The confined energy levels and corresponding wavefunctions were calculated by solving the Schrödinger equation and the Poisson's equation self-consistently using MATLAB. Overlap between electron and hole wavefunctions were obtained by calculating the spatial overlap between the normalized envelop functions. Both TE and TM polarizations were considered in calculation of spontaneous emission rate. The detailed numerical formalism can be found in Ref. [14]. The material parameters of the III-nitride semiconductors were taken from Refs. [90] and [91]. The material parameters of ZnSnN$_2$ were collected from Refs. [24, 37, 43, 44, 68]. The parameter values used in the simulation were summarized in Table 3.1, using the identical notations as in Ref. [14].

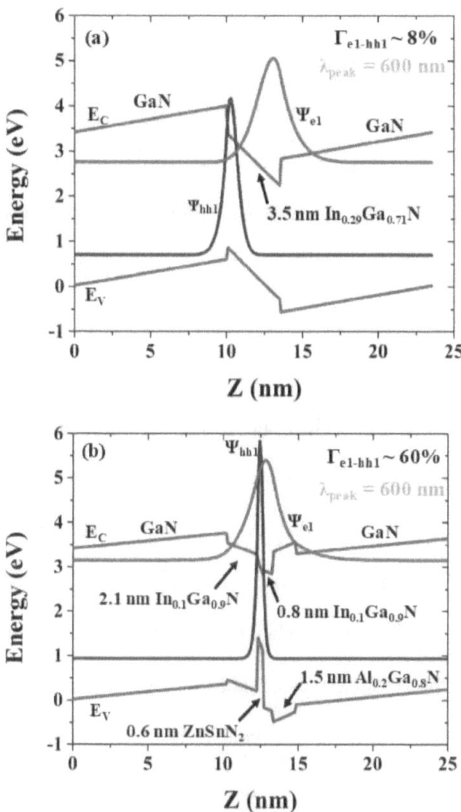

Figure 3.4 Energy-band alignment and electron- and hole-wavefunctions for the first confined energy states in (a) conventional GaN/3.5 nm $In_{0.29}Ga_{0.71}N$/GaN and (b) GaN/2.1 nm $In_{0.1}Ga_{0.9}N$/ 0.6 nm $ZnSnN_2$/ 0.8 nm $In_{0.1}Ga_{0.9}N$/1.5 nm $Al_{0.2}Ga_{0.8}N$/GaN QW. Both QWs were designed with ~600 nm peak spontaneous emission wavelength. [35]

3.4 Energy band alignment

The band edge alignment of the conduction and valence bands for the conventional 3.5 nm $In_{0.29}Ga_{0.71}N$ QW and the 2.1 nm $In_{0.1}Ga_{0.9}N$-0.6 nm $ZnSnN_2$-0.8 nm $In_{0.1}Ga_{0.9}N$-1.5 nm $Al_{0.2}Ga_{0.8}N$ QW are plotted in Figs. 3.4(a) and 3.4(b), respectively. Both structures were designed with peak emission wavelength of ~600 nm. The corresponding electron wavefunction (Ψ_{e1}) and hole wavefunction (Ψ_{hh1}) in the first confined conduction energy states are also plotted. As shown in Fig. 3.4(a), the electron and hole wavefunctions are spatially separated in the IGN QW due to the severe band bending. Consequently, the electron-hole wavefunction overlap Γ_{e1-hh1} is only 8%. On the other hand, the IGN-ZTN QW as shown in Fig. 3.4(b) shows a strong hole wavefunction confinement and a significantly enhanced electron-hole wavefunction overlap of 60%.

It is worth noting that the In-content used in the IGN-ZTN QW (10%) is significantly lower than that used in the conventional IGN QW (29%). Experimentally, growth of higher-In content InGaN is challenging due to the requirement of lower growth temperature for incorporation of more indium, which often leads to reduced material quality associated with severe nonradiative recombinations in LEDs [93, 94, 95]. Therefore, one can expect the crystalline quality of the proposed IGN-ZTN QW structure to be superior to the conventional IGN QW, since higher growth temperature can be implemented for the novel structure.

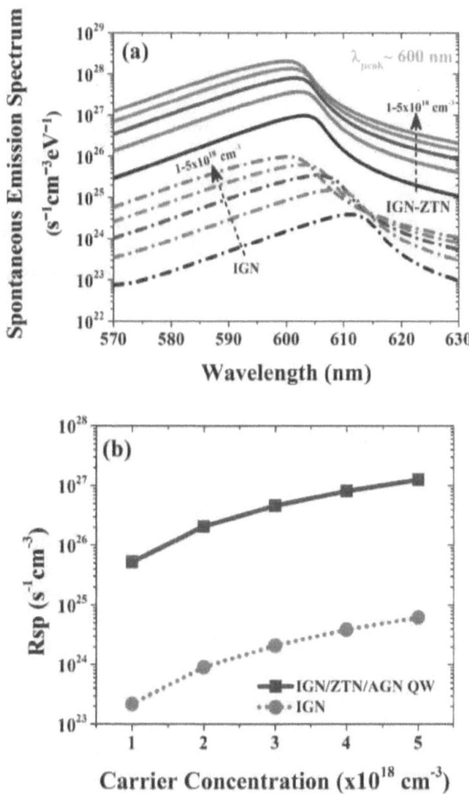

Figure 3.5 (a) Spontaneous emission spectra and (b) spontaneous emission radiative recombination rates of 3.5nm $In_{0.29}Ga_{0.71}N$ (dash-dotted lines denoted by IGN) and 2.1 nm $In_{0.1}Ga_{0.9}N$/ 0.6 nm $ZnSnN_2$/0.8 nm $In_{0.1}Ga_{0.9}N$/1.5 nm $Al_{0.2}Ga_{0.8}N$ (solid lines denoted by IGN-ZTN) QWs for carrier concentrations $1-5\times10^{18}$ cm⁻³. Both QWs were designed with 600 nm peak emission wavelength at 5×10^{18} cm⁻³ carrier concentration. [35]

3.5 Spontaneous emission properties

The spontaneous emission spectra of the IGN-ZTN QW were calculated for carrier concentrations at 1-5×10^{18} cm^{-3}, as compared to that of the conventional IGN QW. As shown in Fig. 3.5(a), both QW structures show a peak emission wavelength of ~ 600 nm at the carrier density of 5×10^{18} cm^{-3}. The peak spontaneous emission intensity I_{peak} for the IGN QW increases from 4.0×10^{24} s^{-1}cm^{-3}eV^{-1} to 9.9×10^{25} s^{-1}cm^{-3}eV^{-1} with the increase in carrier concentration from 1×10^{18} cm^{-3} to 5×10^{18} cm^{-3}. The I_{peak} of IGN-ZTN QW increases from 1×10^{27} s^{-1}cm^{-3}eV^{-1} to 2.1×10^{28} s^{-1}cm^{-3}eV^{-1}, corresponding to approximately 210-250X enhancement. Based on the Fermi's Golden rule, the enhancement in I_{peak} is attributed to the significantly increased electron-hole wavefunction overlap in the IGN-ZTN QW active region. Furthermore, both spontaneous emission spectra sets show blue shifts as the carrier concentration increases due to the carrier screening effect. However, the blue shift of λ_{peak} for the IGN-ZTN QW (~3 nm) is much suppressed as compared to that of the IGN QW (~11 nm). This indicates that the effective band bending in the novel QW design is much reduced as compared to the conventional IGN QW.

The spontaneous emission radiative recombination rate per unit volume R_{sp} is calculated by integrating the spontaneous emission spectrum over the entire wavelength range. As shown in Fig. 3.5(b), R_{sp} increases monotonically with the increase in carrier concentration for both QW structures. The IGN-ZTN QW provides 210-235X enhancement of R_{sp} as compared to that of the IGN QW. Specifically, the R_{sp} of the IGN-ZTN QW increases from 5.2×10^{25} s^{-1}cm^{-3} to 1.3×10^{27} s^{-1}cm^{-33} for carrier concentration at 1-5×10^{18} cm^{-3}, whereas the R_{sp} of the IGN QW is limited to 2.2×10^{23} s^{-1}cm^{-3} - 6.2×10^{24} s^{-}

^{1}cm^{-3}. Note that the internal quantum efficiency (IQE) of LEDs is determined by the ratio of radiative recombination rate and the total recombination rate which includes both the radiative and nonradiative components. Here, if we take into account the expected lower nonradiative recombination in the IGN-ZTN QW, one can expect even larger enhancement of the IQE from the novel QW design.

3.6 Conclusions

In conclusion, a novel QW design for amber LEDs using InGaN-ZnSnN$_2$ QW active layer was investigated. The ZnSnN$_2$ layer inserted in InGaN QW serves as a strong confinement layer for hole wavefunctions due to the large valence band offset. The close to zero conduction band offset between InGaN and ZnSnN$_2$ offers additional advantages for achieving high electron-hole wavefunction overlap. As a result, the peak spontaneous emission intensity and the spontaneous emission radiative recombination rate of the InGaN-ZnSnN$_2$ QW have shown 210-250X and 210-235X enhancement as compared to those of the conventional InGaN QW. In addition, by utilizing the smaller band gap of ZnSnN$_2$ and the large valence band offset with InGaN, only low In-content InGaN is required to achieve amber emission. The proposed IGN-ZTN QW structure is promising to address the current challenge of low efficiency in InGaN QW LEDs emitting beyond blue and green. This work has a great potential to pave a new way to realize high performance monolithic III-nitride LEDs emitting in the entire visible wavelength regime.

Chapter 4

MOCVD growth of ZnGeN$_2$ films on sapphire

4.1 Introduction

The development of the growth of ZnGeN$_2$ and the understanding of its fundamental properties are still at a relatively early stage. There have been only a handful of reports on the growth of ZnGeN$_2$ thin films. These include halide vapor phase epitaxy growth on sapphire [51], MOCVD growth on sapphire with (0001) c- [56, 59], (11-20) r- [57, 56, 59], and (10-12) a- [59] out-of-plane orientations as well as on GaN/c-sapphire templates [60], and MBE growth on GaN/c-sapphire templates [58]. MOCVD growth of $(ZnGe)_{1-x}Ga_{2x}N_2$ thin films on c- and r-plane sapphire as well as GaN/c-sapphire was also reported [61].

This chapter presents a systematic study of the growth of ZnGeN$_2$ by MOCVD on c-, r-, and a-plane sapphire substrates. 2θ-ω X-ray diffraction (XRD) spectra show single crystal films with growth directions consistent with orientations along the orthorhombic Pna2$_1$ [001] direction for c- and a-sapphire substrates, and along the [010] direction for r-sapphire substrates. Both the Zn/Ge atomic percentage ratios and the growth rates were observed to decrease with growth temperature. Room temperature (RT) photoluminescence (PL) peaks observed at 2.05 eV correspond to deep level defect transitions, similar to those often observed for GaN. RT PL excitation (PLE) peaks observed at ~ 3.4 eV are consistent with the theoretically predicted band gap of $ZnGeN_2$ [43] and with previous reported measurements of the band gap by PL [27, 39]. The as-grown $ZnGeN_2$ films were found to be unintentionally doped with electron concentrations of $2x10^{18} - 2x10^{19}$ cm^{-3} that decreased with growth temperature. In addition, effects of thermal on $ZnGeN_2$ grown on c-sapphire substrates were investigated.

The results in section 4.2 have been published in Ref. [59].

4.2 MOCVD growth of ZnGeN₂ films on sapphire substrates with different orientations

4.2.1 Experimental details

The samples in this work were grown in the reactor 2 (R2) of the nitride MOCVD system (CVD03) located at the Nanotech West Lab of the Ohio State University. The $ZnGeN_2$ films were grown on c-, r- and a-plane sapphire substrates using the precursors of diethylzinc (DEZn), germane (GeH_4) and ammonia (NH_3) and the carrier gas N_2. The growth temperature was varied from 600 °C to 710 °C. The DEZn/GeH_4 molar flow rate ratio (II/IV ratio) was varied from 10 to 25 while the NH_3/(DEZn+GeH_4) flow rate ratio (V/(II+IV) ratio) was kept at ~1650, and the reactor pressure at 500 torr. The substrates were *ex situ* cleaned with acetone and isopropanol sequentially, rinsed with deionized water, and blown dry using high purity N_2 prior to insertion into the growth chamber. High temperature (900 °C) *in situ* cleaning with N_2 was employed for 5 minutes prior to initiation of the growth.

XRD spectroscopy was done using a Bruker D8 Discover XRD (Cu K$_\alpha$ source, wavelength 1.5406 Å). The Zn/Ge atomic ratios were determined by energy-dispersive X-ray spectroscopy (EDS). The surface morphologies were characterized by field emission scanning electron microscopy (FESEM). Film thicknesses and thus growth rates were obtained from cross-sectional FESEM images. A Helios Nanolab 600 was used for both EDS and FESEM imaging. Surface roughnesses of the films were obtained from atomic force microscope (AFM) images using a Bruker Icon 3. The optical properties of the as-grown films were probed by RT PL and PLE spectroscopy using a Horiba Fluorolog 3 system equipped with single photon counting, and a Xenon lamp. Carrier concentrations and mobilities were measured at room temperature by the van der Pauw method using an Ecopia HMS-3000 instrument equipped with a magnetic field of 0.975 T.

4.2.2 Crystalline, morphological, optical and electrical properties

The XRD 2θ-ω spectra of the ZnGeN$_2$ films grown on c-, r- and a-sapphire substrates, respectively, are shown in Fig. 4.1 (a)-(c), with II/IV ratios and substrate temperatures of 17 and 680 °C, 25 and 620 °C, and 25 and 680 °C, respectively. The XRD spectrum of the sample grown on c-sapphire (Fig. 4.1(a)) shows two peaks attributed to ZnGeN$_2$, at 2θ ~34.7° and 2θ ~73.2°. These peaks correspond to the (002) and (004) crystal planes of orthorhombic ZnGeN$_2$ [27, 57, 96], indicating that the film growth direction is along the [001] axis. The XRD 2θ-ω spectrum of the film on r-sapphire (Fig. 4.1(b)) showed only one peak, from ZnGeN$_2$, at 2θ~57.8°. This peak, corresponding to the (040) plane of orthorhombic ZnGeN$_2$ [57, 97], shows that the growth direction is along [010], as was observed in Ref. [57]. For the film grown on an a-sapphire substrate, the XRD 2θ-

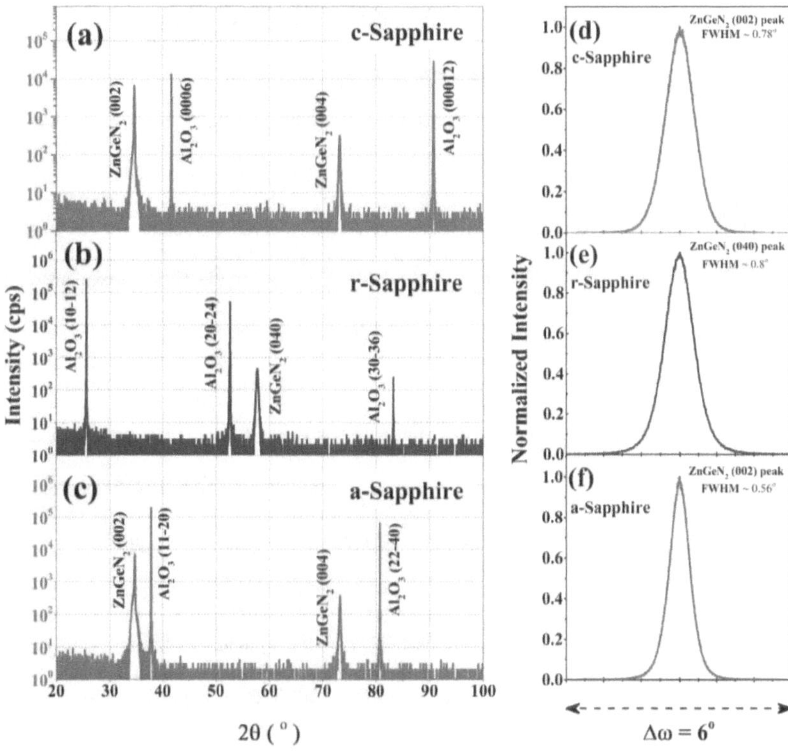

Figure 4.1 (a-c) XRD 2θ-ω spectra and (d-f) XRD ω-rocking curves around symmetric ZnGeN2 peaks measured from three representative ZnGeN$_2$ films grown on c-sapphire (a, d), r-sapphire (b, e) and a-sapphire (c, f) substrates. The group II/IV ratio and growth temperature for each case were (a, d) 17 and 680 °C, (b, e) 25 and 620 °C, and (c, f) 25 and 680 °C. [59]

ω spectrum, shown in Fig. 4.1(c), indicates that the growth direction is along [001], as is the case for the growths on c-sapphire substrates. Therefore, the growth directions of $ZnGeN_2$ on c-, r-, and a-sapphire are similar to those on GaN on respective substrates. The absence of any other peaks corresponding to secondary phases such as Zn_3N_2 or Ge_3N_4 indicates that the grown films are pure $ZnGeN_2$. XRD ω-rocking curves were measured around the $ZnGeN_2$ (002), (040) and (002) peaks for the films grown on c-, r- and a-sapphire, respectively, with measured full widths at half maxima (FWHM) of 0.78°, 0.80° and 0.56° as shown in Fig. 4.1(d-f), respectively. The narrower FWHM for the film grown on a-sapphire may be due to the smaller lattice mismatch between $ZnGeN_2$ and a-sapphire. In case of $ZnGeN_2$ films on c-sapphire, the lattice mismatch between the $ZnGeN_2$ (001) and the sapphire (0001) planes is ~16 % [56]. For $ZnGeN_2$ films grown on r-sapphire, the lattice mismatch with the substrate is ~13 % along the $ZnGeN_2$ [100] and 1.2 % along the $ZnGeN_2$ [001] in-plane directions [57]. For the a-plane sapphire substrate, the in-plane lattice mismatch is 1.6 % along the sapphire [0001] direction and along [1 -1 0 0] is 0.6 % [96].

Deviations from stoichiometry affect the electronic and optical properties of the heterovalent ternary compounds [45, 98, 99]. We find that the Zn/Ge atomic ratio depends strongly on the growth temperature, the ratio of the group II/IV precursors, and the chamber pressure. As the growth temperature increases, the Zn/Ge atomic ratio decreases. An increase in chamber pressure promotes the incorporation of Zn. Figure 4.2 illustrates the effect of growth temperature on the Zn/Ge atomic ratio in the films grown on the three

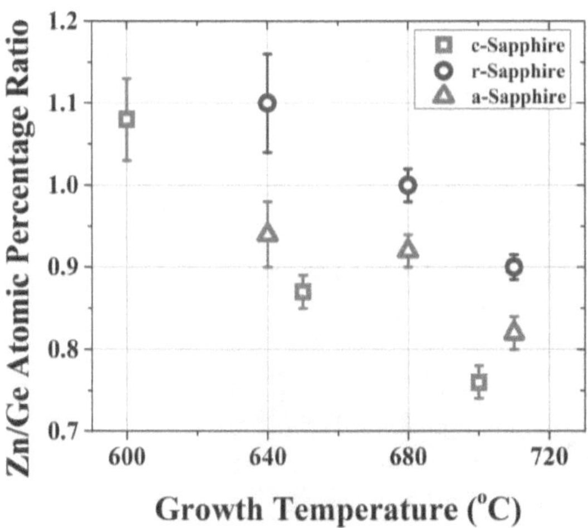

Figure 4.2 Zn/Ge atomic percentage ratio of ZnGeN$_2$ films grown on c-, r- and a-sapphire substrates as a function of the growth temperature. The group II/IV ratio was 10 for the film grown on c-sapphire and 25 for the films grown on r- and a-sapphire. [59]

different sapphire substrates. The group II/IV molar ratio of 10 was used for the films grown on c-sapphire, and 25 for the films grown on r-sapphire and a-sapphire. In Fig. 4.3, it is seen that the Zn/Ge atomic ratios in all films decrease as the growth temperature increases, regardless of the substrate orientation. This trend is attributed to the temperature dependence of the sticking coefficient, or fraction of the atoms incident on the growth surface that are incorporated into the crystal, of the Zn adatoms as compared to that of the Ge adatoms [100], or, a consequence of the much higher vapor pressure of Zn. (The vapor pressure of Zn is approximately 10 orders of magnitude higher than that of Ge at these growth temperatures [101].) The strong dependence of the Zn incorporation on temperature

narrows the window for the growth of stoichiometric ZnGeN$_2$ by MOCVD.

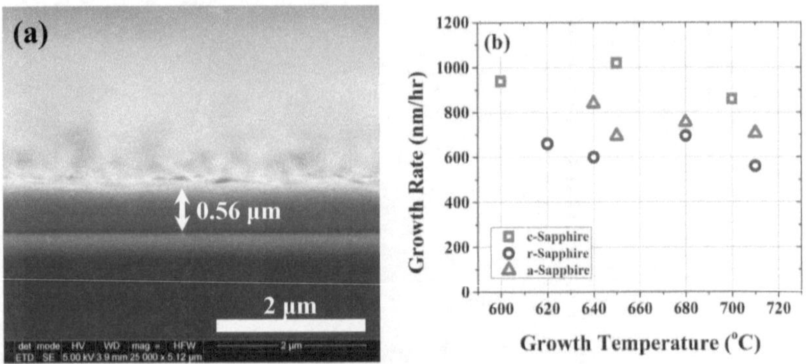

Figure 4.3 (a) Cross-sectional FESEM image of a representative ZnGeN$_2$ film grown on c-

sapphire with a group II/IV ratio of 10 and growth temperature of 650 °C. (b) The film

growth rate versus growth temperature for ZnGeN$_2$ films grown on three types of sapphire

substrates. The group II/IV ratio was 10 for the films on c-sapphire and 25 for the films on

r- and a-sapphire. [59]

Figure 4.3(a) shows a representative cross-sectional FESEM image of a ZnGeN$_2$

film grown on a c-sapphire substrate. Figure 4.3(b) shows the growth rate of the ZnGeN$_2$

films grown on the three types of sapphire substrates as a function of the growth

temperature. The samples grown on c-sapphire substrates used a II/IV ratio of 10, and those

grown simultaneously on r- and a-sapphire substrates used a II/IV ratio of 25. The growth

rates were of the order of 1000 nm/hr, almost an order of magnitude higher than those of

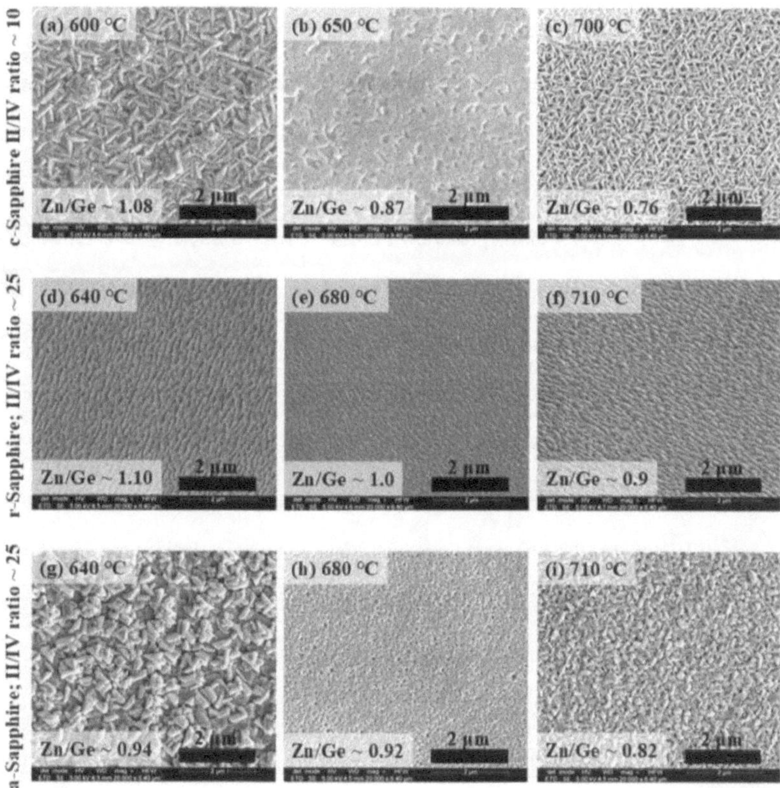

Figure 4.4 Plan view FESEM images of ZnGeN$_2$ films grown on (a-c) c-sapphire, (d-f) r-sapphire and (g-i) a-sapphire at temperatures between 600 °C and 710 °C, with the growth temperatures and measured Zn/Ge atomic ratio shown on each figure. The II/IV flow ratio was 10 for the films grown on c-sapphire substrates (a-c) and 25 for the films grown on r- and a-sapphire substrates. [59]

Ref. [57], and showed a slight decrease, of approximately 150 nm/hr from the lowest to the highest growth temperature used here. Different growth rates for a-, c- and m-planes of GaN, attributed to the differences in surface reaction rates, have also been observed in selective area growth [102, 103].

Figure 4.4 shows plan view FESEM images of $ZnGeN_2$ films grown on c-, r- and a-sapphire substrates under different growth temperatures. Again, the II/IV ratio was 10 for the films on c-sapphire and 25 for those on r- and a-sapphire substrates. The Zn/Ge atomic ratios in these films are shown in Fig. 4.2. For $ZnGeN_2$ grown on c-sapphire substrates, the films grown at 600 °C (Fig. 4.4(a)) and 700 °C (Fig. 4.4(c)) had faceted surfaces, whereas a smoother film was obtained at 650 °C (Fig. 4.4(b)). By increasing the II/IV ratio to 17, continuous films (not shown here) were obtained on c-sapphire substrates at 680°C and 700 °C. This result indicates that the optimum growth temperature for continuous films can be tuned by adjusting the molar flow rates of the cationic precursors. The morphologies of the $ZnGeN_2$ films grown on a-sapphire substrates (Fig. 4.4 (g-i)) showed a similar trend. With a II/IV ratio of 25, they were also smoothest at the intermediate 680 °C growth temperature (Fig. 4.4(h)). The surfaces of the un-faceted films on both c-sapphire and a-sapphire substrates have high densities of pits. On the other hand, the morphologies of the $ZnGeN_2$ films grown on r-sapphire substrates were less sensitive to the growth temperature, compared to the films on c- and a-sapphire substrates. The FESEM images in Fig. 4.4(d-f) show the stepped surfaces of the $ZnGeN_2$/r-sapphire films grown from 640 °C to 710 °C with a II/IV ratio of 25. The directions of the steps are highly oriented.

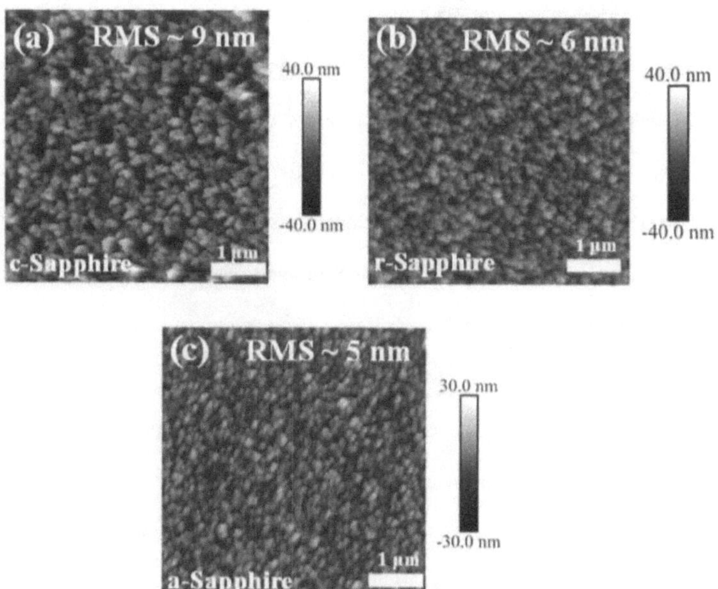

Figure 4.5 5 µm x 5 µm AFM images of the ZnGeN$_2$ films of Fig. 4.1, grown on (a) c-sapphire, (b) r-sapphire and (c) a-sapphire substrates, for which the group II/IV ratios and growth temperatures were (a) 10 and 650 °C, (b) 25 and 640 °C, and (c) 25 and 680 °C. [59]

The surface morphologies of the films were also characterized by AFM over 5 µm x 5 µm scan areas. Figures 4.5(a-c) show the AFM images of the ZnGeN$_2$ films shown in Fig. 4.4(b), 4.4(d) and Fig. 4.4(h), respectively. Large pits on the surface of the ZnGeN$_2$ film grown on c-sapphire are clearly visible in Fig. 4.5(a). For the films grown on r-sapphire substrates, the stepped morphologies seen in the SEM images are also observed in the AFM image Fig. 4.5(b). In addition, this image reveals the existence of triangular pits. The dramatically different surface morphologies between the films grown on r-sapphire and those grown on c- and a-sapphire are related to differences in nucleation processes at the hetero-interfaces on the different substrates. Pits with inverted hexagonal pyramid shapes (V-defects) are a common surface defect of materials that have polarization fields; for example, epitaxial GaN films grown along the polar (0001) direction [102, 104, 105] under kinetically limited conditions. Pits can originate at the meeting point of adjacent islands or at the intersection of a dislocation with the growth surface, where impurities are more likely to accumulate. Clusters of impurities can impede the crystal growth, resulting in the formation of open-core defects bounded by inclined side facets, for example {10-11} planes in the case of (0001) GaN [105]. In the case of non-polar a-plane GaN growth, only half-sides of the V-defects are visible at the surface [102], and the pits appear triangular. Considering the polar nature of the (001) $ZnGeN_2$ surface [68], the inverted pyramidal pits evident on the surfaces of the films on c- and a-sapphire may have a similar origin. The observation of triangular pits on the non-polar (010) $ZnGeN_2$ surface in Fig. 4.5(b) further supports this hypothesis.

Figure 4.6 shows the PL and PLE spectra of the $ZnGeN_2$ films introduced previously, grown on the three types of sapphire substrates. An excitation wavelength of 340 nm was used for PL measurements, and an emission wavelength of 575 nm for the PLE measurements. The spectra for the films grown on the different substrates are spectra that appear similar. The PL spectra are dominated by broad peak at ~ 2.05 eV. This feature has been reported previously, and attributed to deep level defects, and appears similar to the yellow luminescence in GaN [39, 57]. Room temperature PLE peaks observed at ~ 3.4 eV for all samples are interpreted to be a result of exciton enhancement of absorption at the direct band gap. This interpretation is in agreement with the predicted band gap of ordered $ZnGeN_2$ [37] and consistent with reported measurements of the band gap inferred from PL spectra [39, 57].

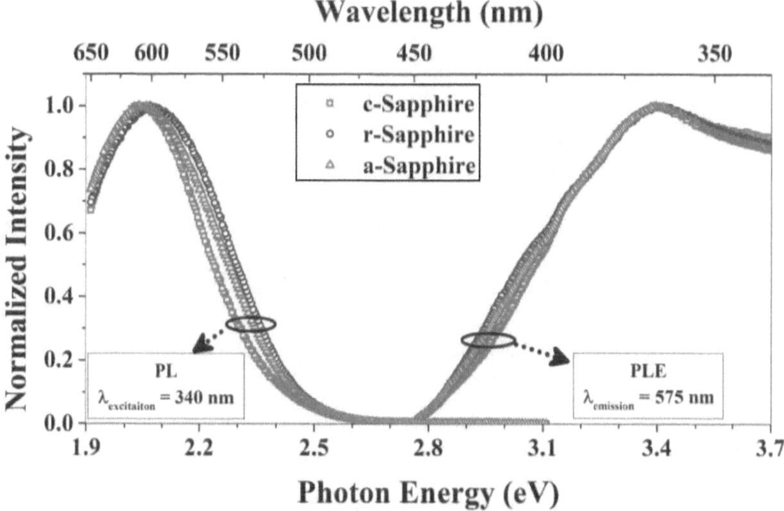

Figure 4.6 PL and PLE spectra of the $ZnGeN_2$ films, of Fig. 4.1 and Fig. 4.6. PL spectra were taken with a 340 nm excitation wavelength. PLE spectra were taken with an emission wavelength of 575 nm, and wavelength resolution of 5 nm. [59]

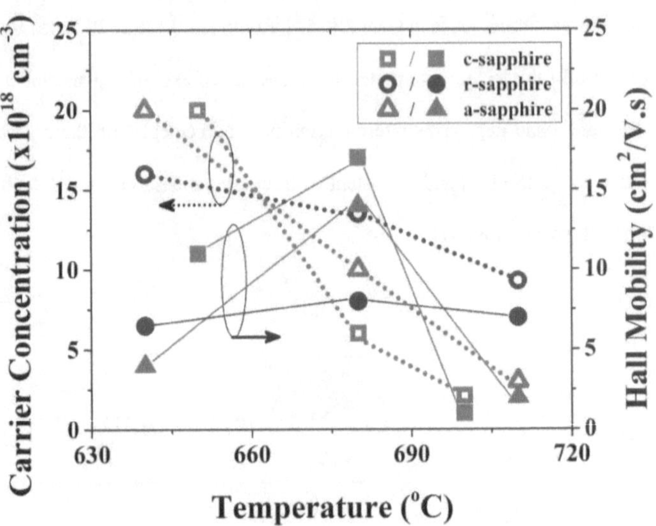

Figure 4.7 Free electron concentrations and mobilities in $ZnGeN_2$ films on c-, r-, and a-sapphire substrates grown at different temperatures measured by van der Pauw Hall measurements. [59]

As-grown $ZnGeN_2$ films were found by room temperature Hall measurements to have unintentional n-type doping. N-type conductivity with electron concentration of 10^{18}-10^{19} cm^{-3} and mobility up to 5 $cm^2/V \cdot s$ in HVPE grown $ZnGeN_2$ films grown on sapphire [51] was previously reported. Figure 4.7 shows the free carrier concentrations (open symbols) and Hall mobilities (solid symbols) for $ZnGeN_2$ films grown on the different sapphire substrate orientations, as functions of the growth temperature. For the films grown on c- and a-sapphire substrates, the electron concentrations were observed to vary from $2x10^{18}$ cm^{-3} to $2x10^{19}$ cm^{-3}. A smaller variation, from 1.0 to 1.6 $\times 10^{19}$ cm^{-3}, was observed for the films grown on r-sapphire substrates. The electron concentrations were observed in all cases to decrease with increases in growth temperatures. First-principles calculations have shown that Ge-at-Zn antisites defects (Ge_{Zn}) and substitutional O on N sites (O_N) are two possible shallow donor species [74]. Native defects such as Zn-at-Ge antisites (Zn_{Ge}) act as acceptors that may compensate the shallow donors [74]. Additionally, H incorporation may contribute to unintentional n-doping in $ZnGeN_2$. The origin of such high unintentional electron concentration from the MOCVD growth still requires further investigation. The room temperature Hall mobility values in the $ZnGeN_2$ films presented in Fig. 4.7 vary from 1 cm²/V·s to 17 cm²/V·s. The Hall mobilities did not vary monotonically with growth temperature, as did the carrier concentrations, but were observed to correlate inversely with the FWHM of the XRD ω-rocking curves. For example, for $ZnGeN_2$ films grown on c-sapphire with higher growth temperature at 680 °C than 650 °C, the measured FWHM value of the rocking curves at (002) peak is also narrower (0.78° vs. 0.96°).

4.3 Thermal annealing of $ZnGeN_2$ films grown on c-sapphire substrate

4.3.1 Experimental details

A $ZnGeN_2$ film was grown on a 2" diameter (0001) sapphire (C-sapphire) substrate via metalorganic chemical vapor deposition using DEZn, GeH_4 and NH_3 as precursors of Zn, Ge and N, respectively. N_2 was used as the carrier gas. Growth temperature, pressure and $DEZn/GeH_4$ molar flow rate ratio were 650 °C, 575 Torr, and 16, respectively. Details of the growth has been described in section 4.2.1 of this book and reported in Ref. [59]. The as grown sample was cleaved radially into pieces. One of the pieces was retained to be used as the as-grown (control) sample and will be referred to as sample A whereas the others were used to investigate different annealing conditions. To prevent the decomposition of the $ZnGeN_2$ film DEZn and NH_3 were flown during the annealing. No GeH_4 was flown during the annealing. Three annealing conditions were investigated, and corresponding samples will, hereafter, be referred to as sample B, sample C, and sample D, respectively. The annealing temperature (T_A), pressure (P_A) and DEZn flow rate (R_{DEZn}) for Sample B were 780 °C, 250 Torr and 54 μmol/min, respectively. Sample C was annealed at $T_A = 800$ °C but under otherwise same conditions as sample B. A $T_A = 800$ °C was also used for sample D but the P_A and R_{DEZn} values were 350 Torr and 81 μmol/min, respectively. These conditions are listed in Table 4.1.

Table 4.1 Temperature (T_A) in °C, pressure (P) in Torr, DEZn and NH_3 molar flow rates in μmol/min, duration (t) in minute used for annealing the samples. The Zn/(Zn+Ge) ratio in the as grown as well as in the annealed samples are also listed.

Sample	Annealing conditions					$\dfrac{Zn}{Zn + Ge}$
	T_A (°C)	P (Torr)	DEZn (μmol/min)	NH_3 (mmol/min)	t (min)	
A (as grown) (ZGN0074)	N/A	N/A	N/A	N/A	N/A	0.47
B (ZGN0074a8)	800	250	54	178	15	0.42
C (ZGN0074a9)	780	250	54	178	15	0.44
D (ZGN0074a10)	800	350	81	178	15	0.45

The atomic percentages of Zn, Ge and N in the as grown as well as the annealed samples were measured by EDS. Surface morphologies of the samples were investigated by FESEM. Both the EDS measurements and the FESEM imaging were carried out using an FEI Apreo LoVac Analytical SEM. Crystallinity and crystalline quality of the samples were investigated using x-ray diffraction measurements using a Bruker D8 Discover XRD with Cu Kα source ($\lambda = 1.5406$ Å). Raman spectra were captured at room temperature using a Renishaw-Smiths Detection Combined Raman-IR Microprobe and 785 nm excitation wavelength. PLE measurements were carried out using a Horiba Fluorolog-3 system equipped with a single photon counter. A Xe arc lamp was used as the excitation source.

4.3.2 Results and discussions

The Zn/(Zn+Ge) cation composition in the samples estimated from EDS data are listed in Table 4.1. Overall, the Zn/(Zn+Ge) composition slightly decreases in the annealed samples as compared to the as grown sample. Among the annealed samples (B-D), the Zn/(Zn+Ge) composition decreases with increase in annealing temperature or decrease in reactor pressure or decrease in DEZn flow rate. This trend is attributed to the much higher equilibrium vapor pressure of Zn as compared to Ge. Therefore, it is essential to flow sufficient amount of Zn and NH_3 precursors during thermal annealing of $ZnGeN_2$ in order to prevent decomposition of the film.

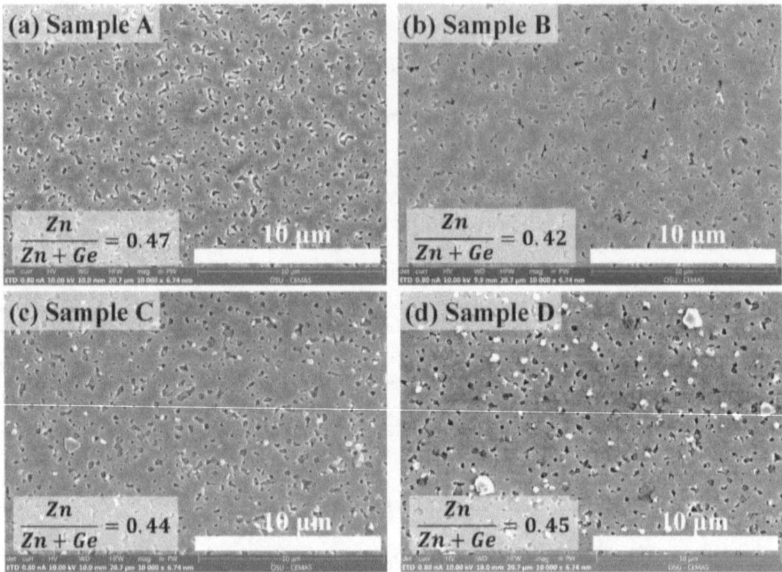

Figure 4.8 Plan-view FESEM images of the (a) as-grown sample A and (b-d) annealed samples B-D, respectively.

Figure 4.8 shows the plan-view FESEM images of the as grown (sample A) as well as three annealed samples (sample B, C, and D). For all four samples, the surface morphology appears to be similar indicating that no significant change was caused by the thermal annealing. A similar surface morphology of the annealed samples as that of the as grown sample also confirms that there was no growth during the annealing process although the DEZn and NH_3 were supplied.

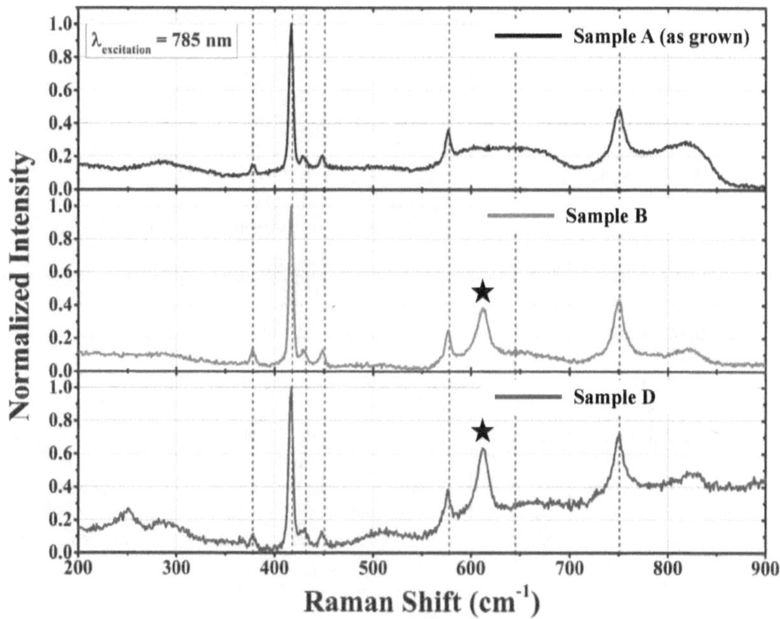

Figure 4.9 Room temperature Raman spectra of as-grown sample A (top), annealed samples B (middle) and D (bottom). Vertical dashed lines mark the position of Raman peaks of c-sapphire. The black stars in the middle and bottom spectra mark the $ZnGeN_2$ related Raman peak in the spectra from annealed samples. Excitation wavelength was 785 nm.

Figure 4.9 presents the room temperature Raman spectra from the annealed samples B and D along with that of the as grown sample A. A 785 nm laser was used as the excitation source. The red vertical dashed lines mark the position of the phonon modes of c-sapphire substrate. $ZnGeN_2$ has 78 phonon modes all of which are Raman active [106].

54

For the as grown sample A (top panel), no peaks related to $ZnGeN_2$ phonon-modes were observed. All the dominant peaks in this spectrum corresponds to the phonon modes in c-sapphire substrates. The only $ZnGeN_2$ related feature observed for this sample was the broad peak near 830 cm^{-1} which corresponds to the phonon density of states (DOS) in the material. The absence of phonon-modes and the appearance of the DOS-like features in the Raman spectrum indicate the presence of cation disorder in the crystal [62]. The Raman spectra of the annealed samples B and D (middle and bottom panel in Fig. 2, respectively) show two prominent differences from that of the as grown sample A – (i) a $ZnGeN_2$ phonon-mode related peak appears at ~612 cm^{-1} and (ii) the $ZnGeN_2$ phonon DOS like features became less prominent. Both of these features indicate a reduced degree of cation disorder in the annealed samples. Thermal annealing induced improvement in cation ordering has previously been reported in Ref. [62].

Figure 4.10(a) shows the room temperature PLE spectra of two annealed samples (B and C) along with that of the grown sample A. The same experimental set up was used for all the samples. The spectra were normalized with respect to the corresponding maximum intensity. For the as grown sample, the PLE spectrum shows a peak at 3.46 eV with a very broad lower energy tail with 1/e of the peak intensity occurring at 3.04 eV. The PLE spectra of the annealed sample C shows a peak at 3.47 eV but with a narrower lower energy tail (1/e of the peak intensity occurs at 3.18 eV) whereas PLE spectra of the annealed sample B shows a peak at 3.51 eV with the narrowest lower energy tail (1/e of peak intensity occurs at 3.29 eV) among all three samples.

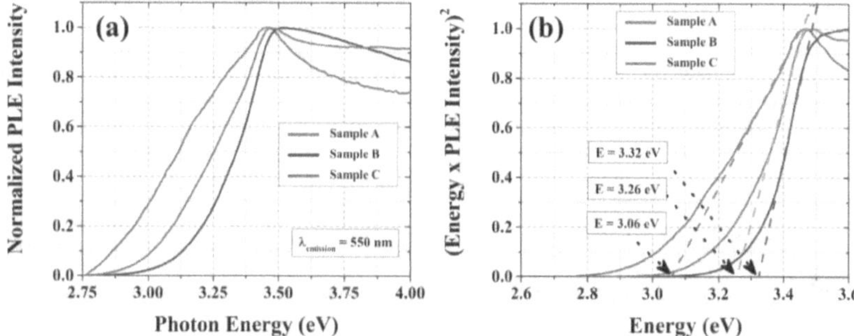

Figure 4.10 (a) Room temperature PLE spectra obtained from the as grown sample A and two annealed samples B, and C. (b) Tauc-like plot constructed by taking the square of the product of PLE intensity by corresponding photon energy values. Straight lines were fitted over the liner region on the lower energy side of the peak for each sample.

Tauc like plots were constructed for sample A, B, and C from the PLE spectra in Fig. 4.10(a) by taking the square of product of the PLE intensity and corresponding photon energy which are shown in Fig. 4.10(b). A line fitting was performed on the tauc-like plot over the linear region one the lower energy side of the peak for each of the sample. The fitted straight lines intercept the energy axis at 3.06 eV, 3.26 eV and 3.32 eV, respectively, for samples A, B, and C indicating an increase in the bandgap from sample A to sample C to sample B. An increase in bandgap of the annealed samples as compared to the as grown sample indicates annealing induced improvement in cation ordering and is consistent with the observation from Raman spectroscopy.

4.4 Conclusions

Heteroepitaxial ZnGeN$_2$ films were grown on c-, r- and a-sapphire substrates via MOCVD. XRD spectroscopy showed that films were single crystal and consistent with either orthorhombic cation ordering or a disordered cation sublattice, and additionally confirmed that the film growth directions were along the c-axis on c- and a-sapphire substrates and along the Pna2$_1$ orthorhombic [010] (wurtzite [11-20]) direction for films grown on r-sapphire substrates. The amount of disorder on the cation sublattice was not determined. For otherwise identical growth conditions, the Zn/Ge atomic percentage ratio in the as-grown films was observed to decrease as the growth temperature increased, as a consequence of the high vapor pressure of Zn. The growth rate of ZnGeN$_2$ films on three different sapphire substrates showed a slight dependence on growth temperature. RT PL spectra with a broad luminescence peak around 2.05 eV indicated the presence of deep level defects, similar to the situation for GaN. RT PLE spectra showed peaks at 3.4 eV, consistent with the theoretically predicted band gap of ordered Pna2$_1$ ZnGeN$_2$. Broad tails in the PLE spectra are consistent with band tailing due to some amount of disorder on the cation sublattice. RT Hall measurements revealed n-type carrier concentrations in the range of $2x10^{18} - 2x10^{19}$ cm^{-3}. Electron concentrations were observed to decrease with an increase in growth temperature. Hall mobilities were in the range of 1 to17 cm^2/V·s. Hall mobilities are expected to increase with further optimization of the film quality. Thermal annealing study was also conducted on ZnGeN$_2$ films grown on c-sapphire substrates. The appearance of ZnGeN$_2$ phonon-mode related peak, which was absent for as grown sample, in the Raman spectra of the annealed samples as well as an increase in the bandgap, estimated from PLE spectra, of the annealed samples as compared to the as grown samples indicated thermal annealing induced enhancement of cation ordering in $ZnGeN_2$.

Chapter 5

Effects of cation stoichiometry on surface morphology and crystallinity

of ZnGeN$_2$ films grown on GaN by MOCVD

5.1 Introduction

In this chapter, we have systematically investigated the MOCVD growth of ZnGeN$_2$ films grown on GaN templates with (0001) orientation. Key growth parameters such as growth temperature, total reactor pressure, and cationic precursor flow rates were mapped to determine their impacts on the cation stoichiometry of ZnGeN$_2$ films grown on closely lattice-matched GaN. Surface morphology and crystalline quality of the films showed strong correlation with the cation stoichiometry. Comprehensive materials characterization indicated that the films are substantially disordered on the cation sublattice.

5.2 MOCVD growth and characterization of ZnGeN$_2$ films on GaN/c-sapphire templates

The ZnGeN$_2$ films were grown in reactor 2 of the nitride MOCVD system at the Nanotech West Lab of the Ohio State University. GaN-on-sapphire templates with (0001) out-of-plane orientation were used as substrates. The substrates were ~3.5-4.5 μm thick and unintentionally doped with free electron concentrations in the range of low 10^{17} cm^{-3}.

59

DEZn, GeH$_4$ and NH$_3$ were used as precursors of Zn, Ge and N, respectively. N$_2$ was used as the carrier gas. Samples were grown at temperatures between 600 °C and 775 °C with reactor pressures between 200 Torr and 500 Torr. The DEZn-to-GeH$_4$ molar flow rate ratio (II/IV ratio, R$_{II/IV}$) ranged between 15 and 75. The NH$_3$ molar flow rate was kept at 178 mmol/min. The substrates were cleaned *ex situ* using acetone and isopropanol, rinsed with deionized water, and blown dry using nitrogen. Prior to epi-growth, the substrates were in-situ annealed for 3 minutes at 900 °C under a combination of N$_2$ and NH$_3$ flow. Total eleven samples were grown for this work. The growth conditions used are listed in Table 5.1.

Table 5.1 Growth temperature (TG) in °C, total reactor pressure (P) in Torr, DEZn/GeH$_4$ molar flow rate ratio (II/IV ratio, R$_{II/IV}$), thickness (d) in μm, Zn/(Zn+Ge) compositions and RMS roughness in nm measured for each labelled sample. The growth temperatures were measured by a thermocouple. The error in Zn and Ge atomic percentages is ~3-4 %. The NH$_3$ molar flow rate was 178 mmol/min in each case. The growth duration was 60 minutes for Samples A-I, 45 minutes for Sample J and 20 minutes for Sample K.

Sample	Growth ID	T$_G$ (°C)	P (Torr)	R$_{II/IV}$	d (μm)	$\frac{Zn}{Zn+Ge}$	RMS roughness (nm)
A	ZGN0106	700	500	25	1.04	0.45	-
B	ZGN0105	650	500	25	1.13	0.50	5.4
C	ZGN0108	600	500	25	1.12	0.54	-
D	ZGN0109	600	350	25	1.10	0.51	-
E	ZGN0111	600	200	25	0.45	0.50	2.7
F	ZGN0118	650	500	20	1.00	0.49	-
G	ZGN0115	650	500	15	0.90	0.45	-
H	ZGN0125	730	500	45	1.40	0.50	7.2
I	ZGN0134	775	500	75	1.22	0.50	-
J	ZGN0087	660	500	25	-	0.47	-
K	ZGN0145	735	500	75	0.40	0.50	2.6

The atomic percentages of Zn, Ge and N in the $ZnGeN_2$ films were measured by EDS. Surface morphologies of the films were investigated by FESEM. The thicknesses of the films were estimated from cross-sectional SEM imaging. An FEI Apreo LoVac Analytical SEM was used for both the EDS measurements and the FESEM imaging. Atom probe tomography (APT) was used to investigate the compositional homogeneity of the cations using a CAMECA Local Electrode Atom Probe (LEAP) 5000 XR system. APT samples were analyzed at a base temperature of 50 K, laser pulse energy of 15 pJ, and an average evaporation rate of 0.005 ions per laser pulse. APT needle-shaped specimens were prepared in an FEI Nova 200 Focused Ion Beam (FIB) using standard lift-out and annular milling techniques. A Bruker Icon 3 AFM was used to measure the roughness of the films. XRD measurements were performed to determine the crystallinity and growth direction of the films using a Bruker D8 Discover XRD with Cu Kα source. Crystalline quality was investigated by scanning tunneling electron microscope (STEM) imaging using a Thermofisher probe-corrected Titan STEM operated at 300 kV. Raman spectra were captured at room temperature using a Renishaw-Smiths Detection Combined Raman-IR Microprobe and 785 nm excitation wavelength. Cathodoluminescence (CL) measurements were performed using a Thermo Fisher Quattro Environmental Scanning Electron Microscope (ESEM) equipped with Horiba H-Clue CL detector. A 325 micron wavelength cw He-Cd laser was used as the excitation source for the PL measurements. The excitation laser power was approximately 1.4 mW. The excitation spot size was approximately 100 microns in diameter. The spectra were resolved using a 0.8 m double monochromator with

2 nm resolution and detected with a Burle C31034A UV-enhanced photon counting cooled

photomultiplier tube.

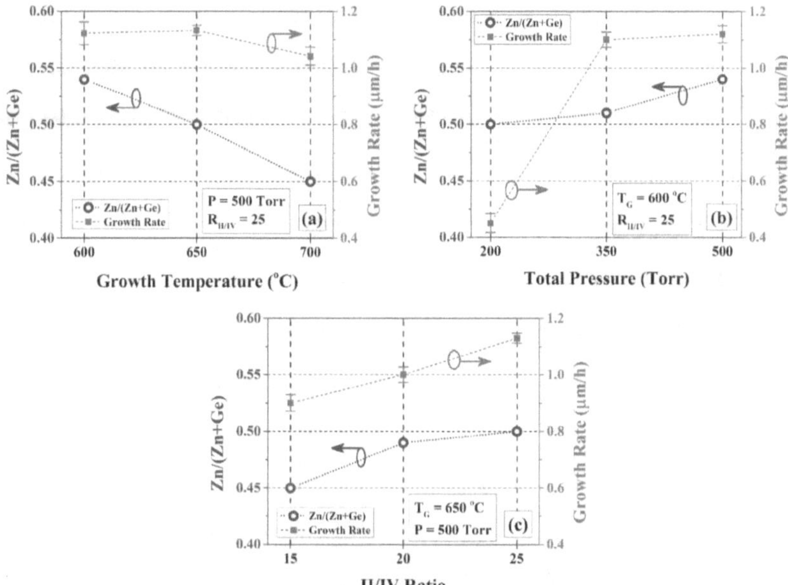

Figure 5.1 The effects of (a) growth temperature T_G (Sample A, B, and C), (b) total reactor

pressure P (Sample C, D, and E), and (c) II/IV ratio $R_{II/IV}$ (Sample B, F, and G) on the

Zn/(Zn+Ge) atomic composition and growth rate of the films. The atomic compositions

were determined from EDS measurements. The error in Zn and Ge atomic percentages is

~3-4 %. [60]

5.3 Effects of growth parameters on ZnGeN$_2$ film properties

5.3.1 ZnGeN$_2$ cation stoichiometry and growth rates

To investigate the effects of MOCVD growth parameters, i.e., growth temperature (T_G), total reactor pressure (P) and group II/IV molar ratio ($R_{II/IV}$) on the stoichiometry and crystallinity of the ZnGeN$_2$ films, three sets of samples were grown while varying only one of these parameters at a time within a set. The total gas flow by volume was kept constant for all growth experiments. The effects of T_G, P, and $R_{II/IV}$ on Zn/(Zn+Ge) composition and film thickness are plotted in Fig. 5.1 (a - c). The cation-to-anion ratios were measured to be stoichiometric, within measurement error. The incorporation rate of Zn into the crystal depends on the concentration of Zn adatoms on the growth surface. The adatom concentration is directly proportional to the pressure of Zn in the vapor phase (P_{Zn}^v) and the flow rate of Zn (J_{Zn}) but inversely proportional to the equilibrium vapor pressure of liquid Zn (P_{Zn}) [107]. For the Samples C, B, and A, as shown in Fig. 5.1 (a), the Zn/(Zn+Ge) composition monotonically decreased from 0.54 to 0.45 with increase in growth temperature from 600 °C to 700 °C. This decrease is attributed to the effect of the increase in equilibrium vapor pressure of Zn with increase in temperature. For example, the equilibrium vapor pressure of Zn increases to ~75 Torr at 700 °C from ~15 Torr at 600 °C [101]. The vapor pressure of Ge is approximately 11 orders of magnitude less than that of Zn over this temperature range. For the Samples E, D and C, as shown in Fig. 5.1 (b), the Zn/(Zn+Ge) composition increased from 0.5 to 0.54 with increase in total reactor pressure from 200 Torr to 500 Torr. This effect can be ascribed to the increase in the pressure of Zn in the vapor phase (P_{Zn}^v) with an increase in the total reactor pressure (P). Finally, the increase in II/IV ratio from 15 to 25 (Samples G, F, and B, respectively) led to the increase

of the Zn/(Zn+Ge) composition from 0.45 to 0.5, as expected (Fig. 5.1(c)).

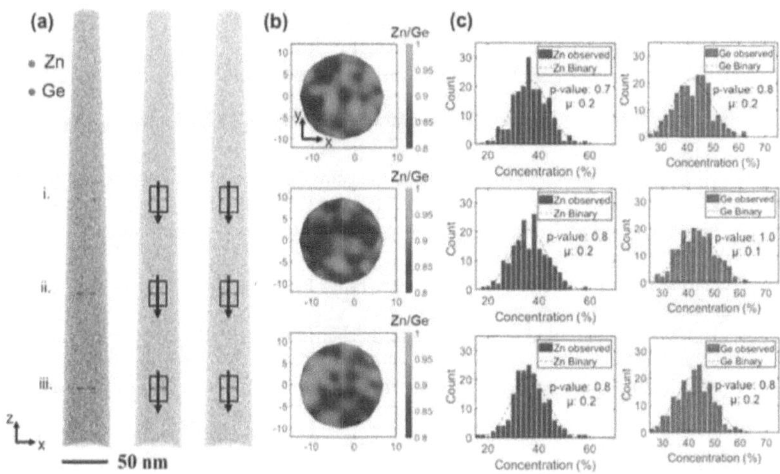

Figure 5.2 (a) Three-dimensional atom maps of Zn (violet) and Ge (green) from APT

analysis of Sample J (from left to right: Zn and Ge, Zn alone, and Ge alone). Three regions

of interest are labelled i-iii, and are shown as blue cylinders with diameters of 20 nm and

thickness of 3 nm. These regions were selected to generate (b) two-dimensional contour

plots (2DCPs) of the Zn to Ge ratio. (c) Frequency of the experimental and calculated

binomial distributions for Zn and Ge are plotted with the inset displaying the respective

Pearson coefficient (μ) and probability-value (p-value). The regions of interest for the

frequency distribution plots are shown as black boxes on the atom maps and have

dimensions of 20 x 15 x 2 nm. [60]

APT measurements were performed on a $ZnGeN_2$ film grown at $T_G = 660$ °C with $P = 500$ Torr and $R_{II/IV} = 25$ (Sample J). The Zn/(Zn+Ge) composition measured by APT was found to be 0.47. APT provided insight on the compositional homogeneity of Zn and Ge atoms across the cross section of the $ZnGeN_2$ films. 3D atom maps of Zn and Ge from APT data are shown in Fig. 5.2 (a). Zn and Ge atoms are shown in violet and green dots, respectively, in the atom maps. Regions of interest are shown in Fig. 5.2 (a) as blue cylindrical slices, labelled i-iii. These measure 20 nm in diameter in the x-y plane and 3 nm in thickness along the z axis. These regions were selected to generate two-dimensional contour plots (2DCPs) of the Zn-to-Ge ratio, shown in Fig. 5.2 (b). Assessment of the 2DCPs revealed relatively homogenous distributions with a range of 0.8-1.0 Zn/Ge. Statistical analyses were performed within regions of interest, boxed in black in Fig. 5.2 (a), measuring 20 x 15 nm in the xz plane and 2 nm thick, to define the likelihood of segregation or clustering of Zn and Ge. Frequency distribution analyses (FDA) allowed for comparison of the experimental distribution and a calculated binomial distribution for each species. The resulting distributions are shown in Fig. 5.2(c) along with statistical measurements of the Pearson coefficient (μ) and the probability-value (p-value). The binomial and experimental distributions are close, indicating a nearly random distribution. The μ reported here is independent of sample size and is analogous to a goodness of fit, for which a value close to 1 indicates a complete association in the occurrence of the solute atoms and a value close to 0 indicates a random distribution. The p-value is a measure of confidence. A p-value of 0.01 indicates a confidence level of 99%. The p-value is not independent of sample size, and therefore μ is the more reliable indicator for segregation.

Both Zn and Ge demonstrate high p-values (close to 1) and low μ values, which indicate a lack of segregation or inhomogeneity. The combined analyses indicate a homogeneous distribution of Zn and Ge. Thus, there is no indication of the presence of a secondary phase, such as Zn_3N_2 or Ge_3N_4.

The growth rates of the $ZnGeN_2$ films were determined from cross-sectional SEM imaging. The measured values were in agreement with the estimated growth rates from the *in situ* reflectivity monitor. The error bars arise from the uncertainty due to the surface roughness plus the small variation in the thickness from the edge towards the center of the 2" wafer. With 500 Torr total pressure and a II/IV ratio of 25 (Sample A-C), the growth rates of the films were close to 1.1 μm/hr. These growth rates did not vary significantly with growth temperature (Fig. 5.1(a)). At 600 °C, the growth rate remained almost constant (~1.1 μm/hr) as the total pressure decreased from 500 Torr to 350 Torr (Sample D) but dropped to ~0.45 μm/hr at 200 Torr (Sample E) (Fig. 5.1(b)). On the other hand, for T_G = 650 °C and P = 500 Torr the growth rate increased monotonically, from approximately 0.9 to 1.1 μm/hr with an increase in the II/IV ratio from 15 to 25 (Sample G, F, and B).

5.3.2 ZnGeN₂ surface morphologies and growth conditions

The plan-view FESEM images of the films presented in Fig. 5.1 are shown in Fig. 5.3. The surface morphologies of the films grown under otherwise identical conditions (P = 500 Torr, $R_{II/IV}$ = 25) but at T_G = 700 °C (Sample A), 650 °C (Sample B) and 600 °C (Sample C) are plotted in Fig. 5.3(a - c), respectively. Figures 5.3(d) and 5.3(f) show the films grown at T_G = 650 °C and P = 500 Torr (same as the Sample B in Fig. 5.3(b)) but

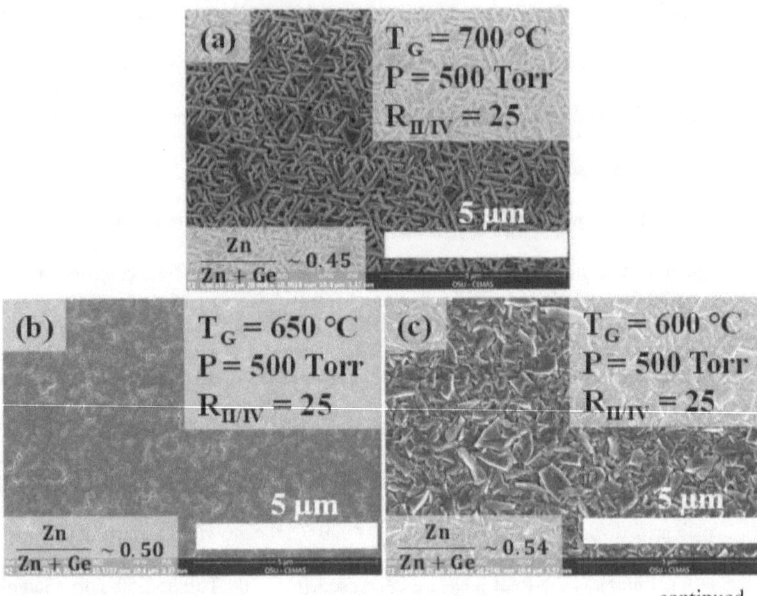

continued

Figure 5.3 Plan view FESEM images of the ZnGeN$_2$ films grown on GaN-on-sapphire templates. Films were grown with (a-c) P = 500 Torr and $R_{II/IV}$ = 25 and T_G = 700 °C (Sample A), 650 °C (Sample B) and 600 °C (Sample C), respectively, (d, f) T_G = 650 °C and P = 500 Torr and $R_{II/IV}$ = 20 (Sample F) and 15 (Sample G), respectively and (e, g) T_G = 600 °C and $R_{II/IV}$ = 25 and P = 350 Torr (Sample D) and 200 Torr (Sample E), respectively. The Zn/(Zn+Ge) compositions in the films in (a-g) are also shown in Fig. 1. The growth parameters {T_G, P and $R_{II/IV}$} for the films in (h) and (i) were {730 °C, 500 Torr, 45} (Sample H), {775 °C, 500 Torr, 75} (Sample I), respectively. [60]

Figure 5.3 continued

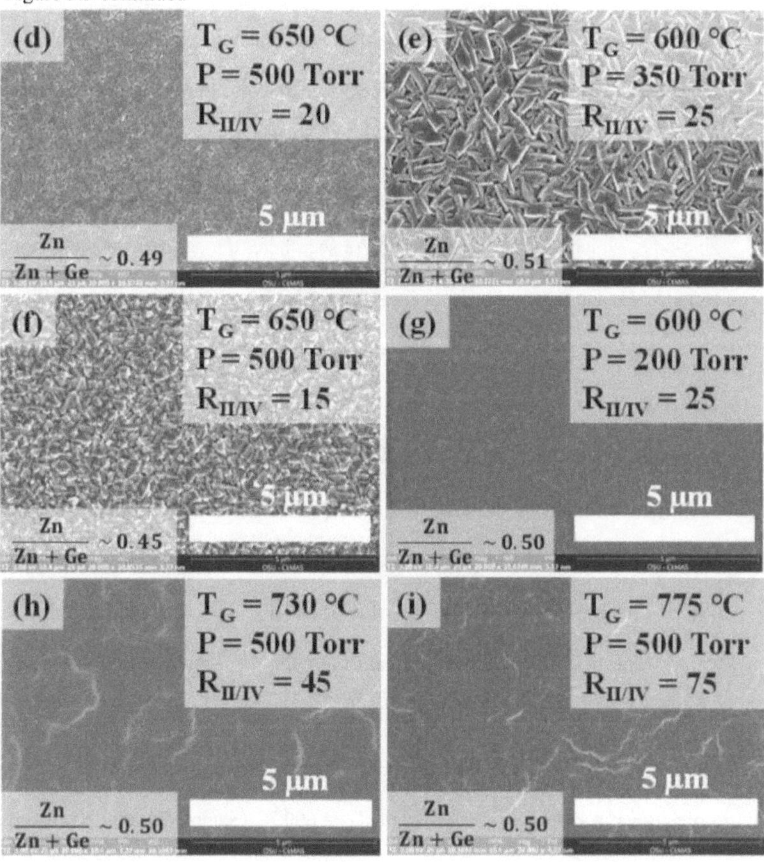

with $R_{II/IV}$ = 20 (Sample F) and 15 (Sample G), respectively. The films shown in Figs. 5.3(e) and 5.3(g), respectively, were grown with chamber pressures of 350 Torr (Sample D) and 200 Torr (Sample E) respectively, but otherwise under the same conditions as the Sample C shown in Fig. 5.3(c), with T_G = 600 °C and $R_{II/IV}$ = 25. From Fig. 5.1, it is understood that, depending on the combination of T_G, P and $R_{II/IV}$, the grown film can be zinc-poor, stoichiometric or zinc-rich. As seen here, the surfaces of the zinc-poor films (Figs. 5.3(a) and 5.3(f)) consist of facets aligned along certain preferred directions, and the aspect ratios of the facets vary with growth conditions. In the case of zinc-rich films (Figs. 5.3 (c) and 5.3 (e)), the surfaces consist of crystallites of arbitrary shapes and sizes, without dominant orientations. Finally, the stoichiometric or near-stoichiometric films (Figs. 5.3(b, d, and g)) show planar surfaces, which evolved with growth temperature. Figures 5.3 (h) and (i) are plan view FESEM images of two additional stoichiometric films grown at T_G = 730 °C (Sample H: P = 500 Torr, $R_{II/IV}$ = 45) and T_G = 775 (Sample I: P = 500 Torr, $R_{II/IV}$ = 75), respectively, showing larger features on the surface of the film grown at relatively higher temperatures.

Figures 5.4(a-c) show the 3D images obtained from AFM scans of three stoichiometric $ZnGeN_2$ films grown at T_G = 600 °C (Sample E), 650 °C (Sample B) and 730 °C (Sample H), respectively. The estimated film thicknesses and RMS roughness values of these films were ~0.45 μm and 2.7 nm, ~1.1 μm and 5.4 nm, and ~1.4 μm and 7.2 nm, respectively. Figure 5.4(d) is an AFM image of a stoichiometric film grown at ~735 °C (Sample K) but for only 20 minutes resulting in an estimated thickness of ~0.4

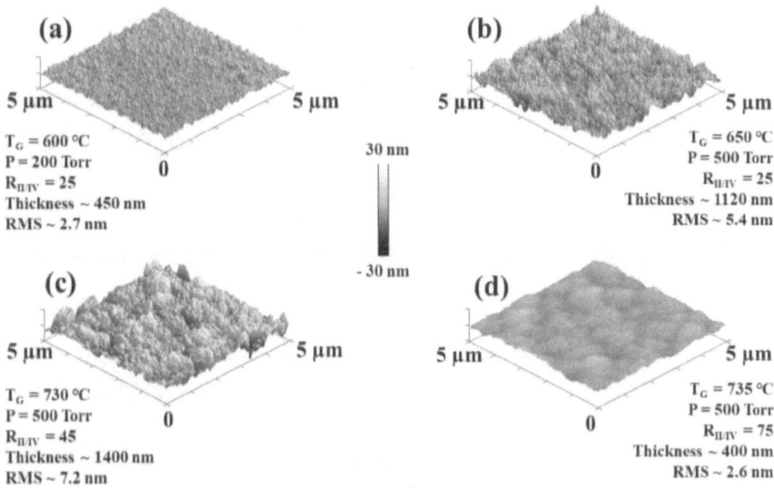

Figure 5.4 3D AFM images (45° rotated, 30° tilted) showing the evolution of the surface of stoichiometric ZnGeN$_2$ films with the increase in growth temperature. Films were grown at (a) 600 °C (Sample E), (b) 650 °C (Sample B) and (c) 730 °C (Sample H) and (d) 735 °C (Sample K), respectively. Thicknesses of the films are 0.45 µm, 1.1 µm, 1.4 µm and 0.4 µm, respectively. Either total reactor pressure or II/IV molar ratio was adjusted to obtain stoichiometric films at different temperatures. The RMS roughnesses of these films are 2.7 nm, 5.4 nm, 7.2 nm, and 2.6 nm, respectively. [60]

µm. The RMS roughness value is ~2.6 nm, which is roughly 1/3 of that of the film grown at 730 °C but for 60 minutes (Fig. 5.4(c)). In general, it was observed that for similar growth temperature, the RMS roughness increases as the film grows thicker. In addition, AFM images in Figs. 5.4(a-c) indicate a gradual change in the dimensions of the surface features with growth temperature. For 600 °C, the film surface (Fig. 5.4(a)) consists of needle-like features with vertical and lateral dimensions in the order of ~10 nm and ~200 nm, respectively. As the growth temperature increases, the feature sizes increase in both vertical and lateral dimensions. For the film grown at 730 °C, the surface is comprised of hillocks with average heights and widths of ~30 nm and 500 nm, respectively (Fig. 5.4(c)). The evolution of the needle-like features to the hillocks is likely caused by the increased adatom mobility at elevated temperatures, facilitating the coalescence of adjacent structures. It is worthwhile noting that such evolution of surface features was observed regardless of the film thickness.

STEM imaging was carried out to better understand the mechanism causing the intriguing dependence of surface morphology of the $ZnGeN_2$ films on Zn/(Zn+Ge) composition. The low angle annular dark field STEM (-LAADF) images of a Zn-rich (Sample C) and a Zn-poor (Sample A) film are shown in Figs. 5.5(a) and 5.5(b), respectively. The cross-section of the Zn-rich film in Fig. 5.5(a) shows a columnar morphology. The columnar growth starts right from the interface between the substrate and the epi-film. Columnar cross-sectional morphology was also observed in the case of polycrystalline $ZnSnN_2$ films grown by combinatorial sputtering for a Zn/(Zn+Sn) composition between 0.45 and 0.7 [45]. On the other hand, the dominant features of the cross section of a Zn-deficient film as shown in Fig. 5.5(b) are tilted filament-like structures, which give rise to the surface morphology shown in Fig. 5.3(a). Under highly Ge-rich growth conditions, the growth of rod-like structures is observed. This phenomenon may be caused by the presence of excess Ge on the growth surface. Figures 5.5(c-e)

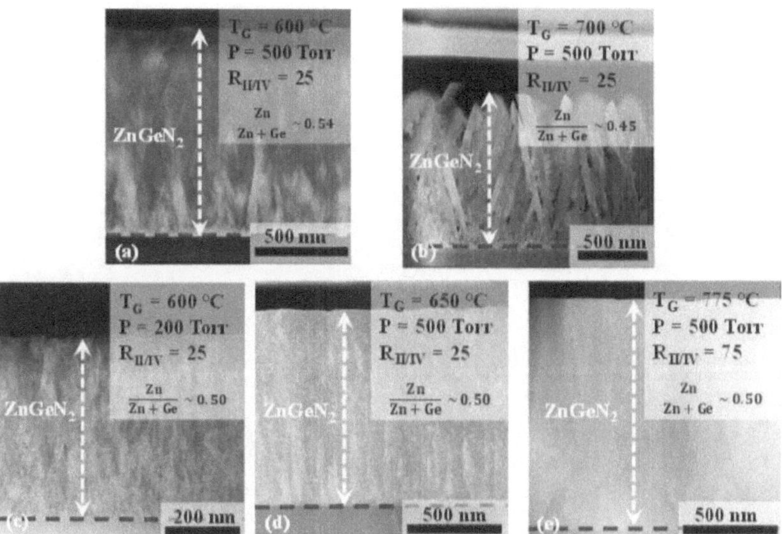

Figure 5.5 STEM-LAADF images showing dramatically different cross-sectional morphology in ZnGeN$_2$ films having different cationic composition. The interfaces between the epi-film and the substrate were marked by blue dashed lines. These films are (a) zinc-rich (Sample C), (b) zinc-poor (Sample A) and (c-e) nominally stoichiometric (Sample E, A and I), respectively. For the samples showed in (c-e), either the total reactor pressure or the II/IV molar ratio was adjusted to obtain a nearly stoichiometric film at each of the different growth temperatures. [60]

represent the cross-sectional STEM-LAADF images of three near-stoichiometric films grown at $T_G = 600\ ^\circ$C (Sample E), 650 $^\circ$C (Sample B) and 775 $^\circ$C (Sample I), respectively. Either the total reactor pressure (P) or the II/IV ratio ($R_{II/IV}$) was varied to obtain near-

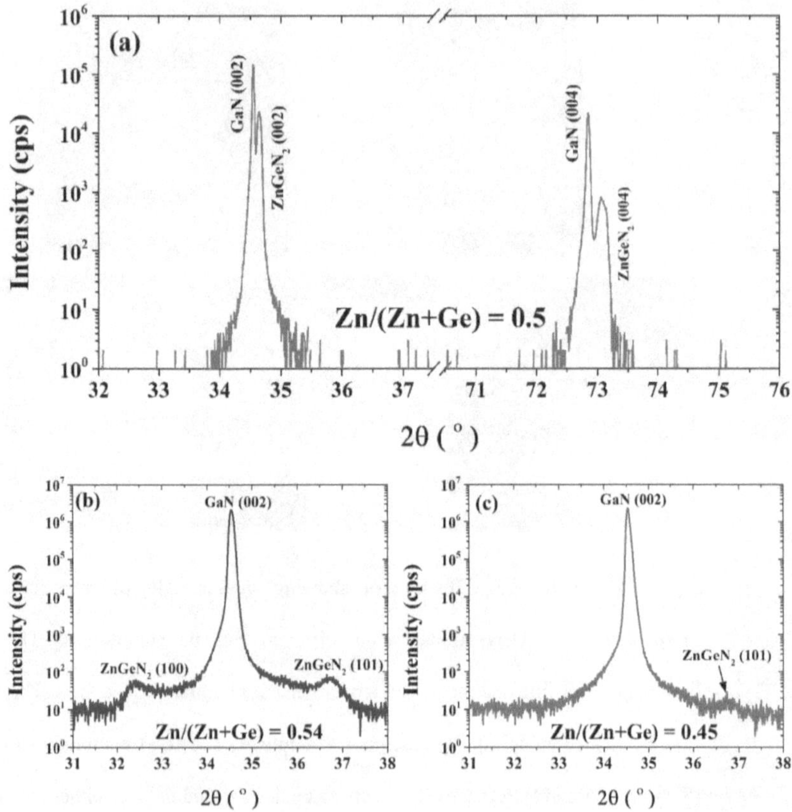

Figure 5.6 XRD 2θ-ω scan-profile from a (a) stoichiometric (Sample B), (b) Zn-rich (Sample C) and (c) Zn-poor (Sample A) ZnGeN₂ film. For the stoichiometric film, only {002} peaks were observed in the range of 2θ = 20° - 90°. The growth parameters {T_G, P and $R_{II/IV}$} were {650 °C, 500 Torr, 25}, {600 °C, 500 Torr, 25} and {700 °C, 500 Torr, 25}, respectively. [60]

stoichiometric cation compositions in these films. The near-stoichiometric films have more uniform cross sections than do the off-stoichiometric films. As the growth temperature increases, the density of extended defects decreases. This effect is likely due to the higher mobility of the adatoms on the growth surface at elevated temperature.

5.3.3 ZnGeN$_2$ crystallinity and crystal structures

Figures 5.6(a-c) show the XRD 2θ-ω scan profiles obtained from a stoichiometric (Sample B), a Zn-rich (Sample C), and a Zn-poor (Sample A) ZnGeN$_2$ film, respectively. In all cases, the highest intensity peak corresponds to the GaN (002) plane. For the stoichiometric film (Fig. 5.6(a)), the XRD 2θ-ω scan profile shows the strong ZnGeN$_2$ (002) peak at 2θ = 34.64° [56, 59], which has comparable intensity to the GaN (002) peak. No other ZnGeN$_2$ peak was observed between 2θ = 20° and 90° except the ZnGeN$_2$ (004) peak at 2θ = 73.07° [56, 59]. However, for the Zn-rich film (Fig. 5.6(b)), two 2θ-ω peaks at 2θ = 32.42° and 36.72° positions are obvious whereas the one at 2θ = 36.72° position is noticeable for Zn-poor film (Fig. 5.6(c)). These two peaks can be assigned to ZnGeN$_2$ (100) and (101) peaks, respectively, assuming a wurtzite structure [97]. It is likely that the intensity of ZnGeN$_2$ (002) peak is low so that it overlaps with the GaN (002) shoulder in Figs. 5.6 (b) and (c). The XRD 2θ-ω scan profiles in Fig. 5.6 indicate that stoichiometric films are single crystalline with (001) out-of-plane orientation, whereas Zn-rich or Zn-poor films are polycrystalline. Moreover, the positions of (002) and (004) peaks indicate that ZnGeN$_2$ has slightly smaller lattice constant than GaN along c-axis.

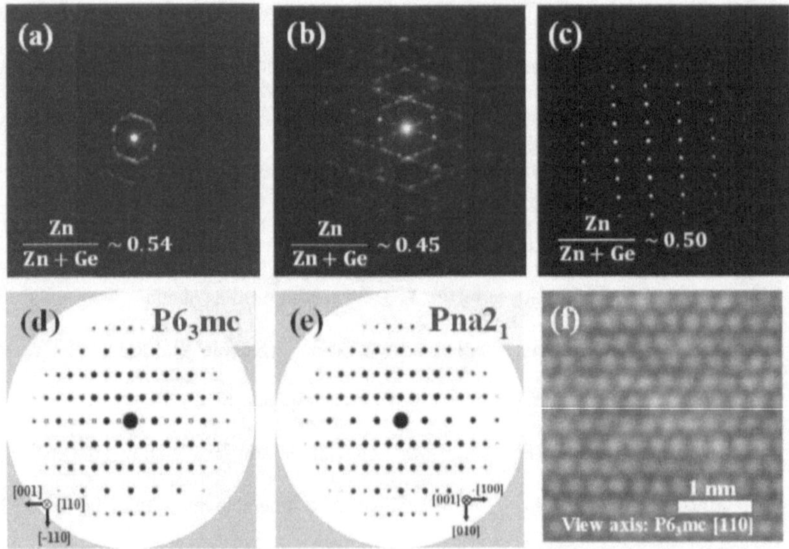

Figure 5.7 Experimental position-averaged nano-diffraction patterns of (a) a Zn-rich (Sample C), (b) a Zn-poor (Sample A), and (c) a stoichiometric (Sample I) $ZnGeN_2$ film, respectively. The growth parameters $\{T_G, P$ and $R_{II/IV}\}$ were {600 °C, 500 Torr, 25}, {700 °C, 500 Torr, 25} and {775 °C, 500 Torr, 75}, respectively. (e, f) Calculated TEM diffraction patterns of $ZnGeN_2$ assuming a P6₃mc (wurtzite) structure and a Pna2₁ (orthorhombic) structure, respectively. (f) High magnification STEM imaging from [110] view direction showing the alternating layers of cations. [60]

It is challenging to distinguish between the ordered (Pna2₁) and disordered (P6₃mc) $ZnGeN_2$ structures grown along the c-direction from the XRD 2θ-ω scan profiles because of the similar Bragg conditions of the primary planes of these polymorphs [27]. The TEM nano-diffraction patterns were used to investigate the crystallinity and crystal structures of

75

the MOCVD grown films. A 15×15 array of nano-diffraction patterns was captured from ~ 830 nm×830 nm area from each investigated sample. In general, for the Zn-rich or Zn-poor films, substantial variations were observed among individual diffraction patterns in the array whereas all 225 patterns were identical for nominally stoichiometric films. Moreover, the stoichiometric films grown under different growth conditions show very similar nano-diffraction patterns. Figures 5.7(a-c) show the average of the 225 nano-diffraction patterns captured from a Zn-rich (Sample C), a Zn-poor (Sample A), and a stoichiometric $ZnGeN_2$ film (Sample I), respectively. These position-averaged diffraction patterns indicate that the near-stoichiometric films are single crystalline whereas the off-stoichiometric films are polycrystalline in structure. In addition, the structural variations in the Zn-rich and Zn-poor films were markedly different, an observation associated with the different cross-sectional morphologies of these films (see Figs. 5.5 (a) and (b)). Figure 5.7(d) and (e) show two simulated electron diffraction patterns calculated assuming completely disordered $P6_3mc$ ([110] view direction) and ordered $Pna2_1$ ([100] view direction) structures, respectively, of $ZnGeN_2$. Hollow squares in (d) mark the space-group absences in the simulated pattern. Neither of these patterns adequately matches the experimental pattern in Fig. 5.7(c) to comment about the degree of disorder in cation sublattice of the grown films. The experimental pattern is consistent with both simulated diffraction patterns, except for the presence of the forbidden peaks, indicated by the open squares in Fig. 5.7(d). These forbidden peaks are also observed in the diffraction pattern of the underlying GaN template (not shown). Therefore, they most likely appear as a result of the inhomogeneous strain caused by the bending of the TEM lamella. Inhomogeneous strain can also be introduced by local variations in the distribution or ordering of the cations or by inhomogeneous stress in the lamella due to the lattice mismatch between the film and the substrate. Other important factors that can result in the appearance of additional diffraction spots are a thicker-than-ideal TEM lamella and amorphous layers on both sides of the lamella, as these conditions can potentially cause dynamical diffraction.

Figure 5.7(f) is a high magnification STEM image showing the alternating cation layers in a stoichiometric $ZnGeN_2$ film. For this lattice orientation there should be equal distributions of Zn and Ge atoms in each cation column for both the fully ordered $Pna2_1$ phase and for wurtzite-like random distributions of the cations. Therefore, we do not expect to obtain information on the degree of cation ordering from such an image. The contrast variation observed here could be a result of the presence of strain, crystalline defects, phase segregation or non-uniform sample thickness.

Figure 5.8 presents the room temperature unpolarized Raman spectrum of a single crystalline stoichiometric $ZnGeN_2$ film (Sample I, red curve), a Zn-rich film (Sample C, blue curve) and a Zn-poor film (Sample A, green curve) along with that of the GaN-on-sapphire template (black curve). A $ZnGeN_2$ unit cell has 78 optical phonon modes, all of which are Raman active [106]. None of these modes are seen here. The three narrow peaks observed in the spectrum are from phonon modes in the GaN template. In the case of the stoichiometric film, two broad features, one between 600 and 700 cm^{-1} and other around 830 cm^{-1} are identified with peaks in the phonon density of states (DOS) in $ZnGeN_2$ [54, 62, 108]. These two features can also be seen, although they are less pronounced, in the spectrum from the Zn-rich film (blue curve). They are absent in the spectrum from the Zn-

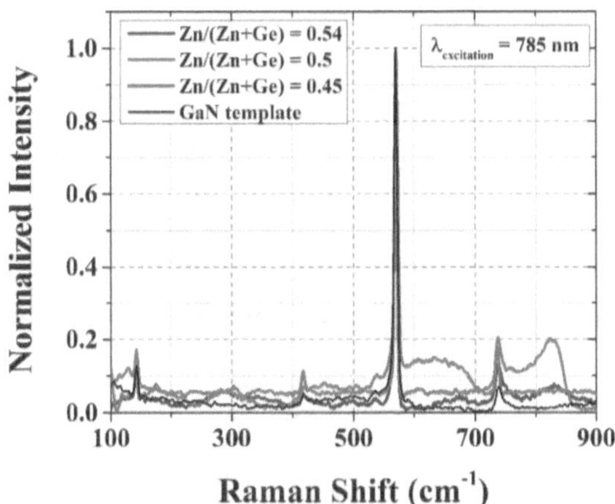

Figure 5.8 Room temperature, unpolarized Raman spectra obtained from a Zn-rich (Sample C, blue curve), a stoichiometric (Sample I, red curve), and a Zn-poor (Sample A, green curve) $ZnGeN_2$ film. The growth parameters {T_G, P and $R_{II/IV}$} were {600 °C, 500 Torr, 25}, {775 °C, 500 Torr, 75} and {700 °C, 500 Torr, 25}, respectively. The Raman spectrum of the GaN-on-sapphire (black curve) is shown for comparison. A 785 nm laser was used as the excitation source. Each spectrum was normalized with respect to the corresponding intensity of the peak at 570 cm^{-1}. [60]

poor film (green curve). Cation disorder in $ZnGeN_2$, evident in XRD pspectra, has been shown to lead to relaxation of the momentum conservation rule and to the corresponding appearance of DOS features in the Raman spectra [27]. The absence of $ZnGeN_2$ Raman peaks in the spectrum in Fig. 5.8 is perhaps not surprising. In Ref. [27], the relative intensities of the Raman versus DOS peaks have been shown to increase for VLS-grown $ZnGeN_2$ on going from a growth temperature of 758 °C to 850 °C, with a

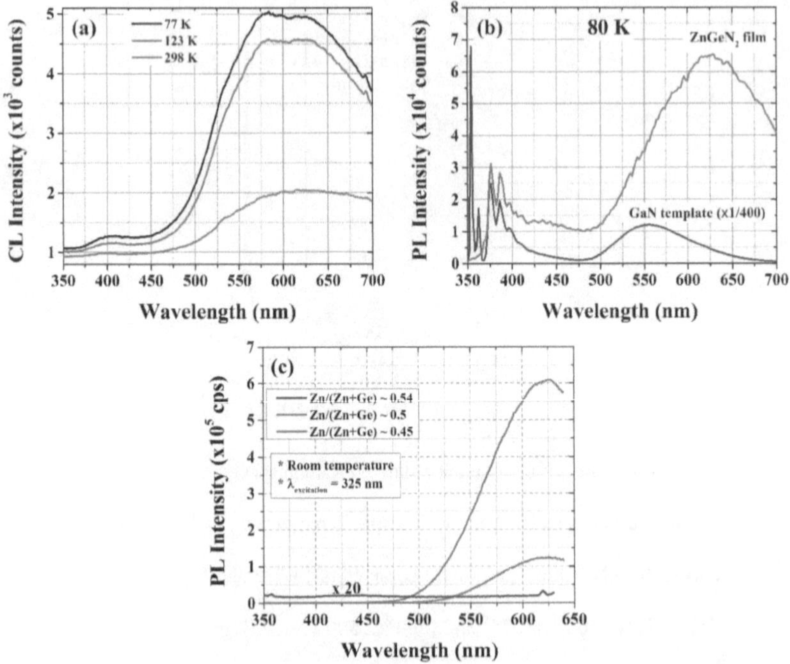

Figure 5.9 (a) CL spectra measured at three different temperatures and (b) PL spectrum measured at 80 K of a stoichiometric $ZnGeN_2$ film grown at $T_G = 730°C$, P = 500 Torr and $R_{II/IV} = 45$ (Sample H). The PL spectrum of a GaN-on-sapphire template measured under the same conditions is shown in (b) for comparison. For the CL measurement, the electron beam energy and beam current were 5 keV and 23 nA, respectively. (c) Room temperature PL spectra measured from a Zn-poor (Sample A), a stoichiometric (Sample I) and a Zn-rich film (Sample C). The intensity of the blue spectrum was multiplied by 20; the actual intensity is similar as the measured background level with identical configuration. [60]

corresponding change in the x-ray powder diffraction scan from disordered (P6₃mc) to ordered (Pna2₁). For the film shown in Fig. 5.8 the growth temperature is 775 °C but the MOCVD growth process may be more kinetically limited than the very different, near-equilibrium VLS process, even for roughly the same growth temperatures.

5.3.4 ZnGeN$_2$ optical properties

Figure 5.9(a) shows the CL spectra of a stoichiometric ZnGeN$_2$ film grown at 730 °C (Sample H) measured at three different temperatures (77 K, 123 K, and 298 K). The electron beam energy and beam current were set as 5 keV and 23 nA, respectively. The sampling area was the same in all cases. No CL peak was observed near the 3.4 eV predicted band gap of orthorhombic ZnGeN$_2$, even at 77 K. At room temperature, a broad peak was observed centered at ~ 615 nm (2.02 eV). At lower temperatures, this broad peak shows clear splitting into two peaks, with one centered at ~560 nm (2.22 eV) and the other at ~640 nm (1.94 eV). The intensities of both peaks increase as the temperature decreases. We attribute these peaks to yellow-band luminescence (YBL) [109] from the underlying GaN template and from the ZnGeN$_2$ film. Similar peaks were observed in room temperature PL spectra of ZnGeN$_2$ films grown on sapphire substrates [59]. Figure 5.9 (b) shows the PL spectrum of the ZnGeN$_2$ film along with the PL spectrum of a GaN-on-sapphire template as a reference. Both spectra were measured at 80 K. From Fig. 5.9(b), it is evident that the peak centered at ~630 nm in the spectrum of the film (red) is primarily associated with defect PL from the ZnGeN$_2$ layer, with possibly some contribution from defect PL from the GaN substrate, whereas the peaks at wavelengths below 400 nm correspond to near-band edge and impurity PL from the GaN substrate. Such "yellow-band-like" defect PL has been observed previously in $ZnGeN_2$ grown by VLS [39] and by MOCVD [57, 59]. Figure 5.9(c) plots room temperature PL spectra collected from a Zn-poor (Sample A), a stoichiometric (Sample I), and a Zn-rich (Sample C) film. The spectra from the Zn-poor and stoichiometric films show broad defect peaks at similar wavelength near 625 nm, but no luminescence from the Zn-rich film was observed. This 'yellow-band-like' luminescence may be related to Ge-at-Zn anti-site or Zn-vacancy type defects [74, 75]. However, further investigation such as defect spectroscopy is still required to confirm the origin of these defect levels.

5.4 Conclusions

In summary, $ZnGeN_2$ films were grown by MOCVD on closely lattice-matched GaN-on-sapphire templates. The stoichiometry of the films was found to be highly dependent on the growth parameters, including growth temperature, total reactor pressure and the group II/IV molar ratio. This work has demonstrated that stoichiometric $ZnGeN_2$ films on GaN can be achieved in a wide growth window by tuning combinations of these key parameters. The crystallinity and surface morphologies of the films were found to be highly dependent on the stoichiometry. APT measurements indicated that the local distribution of cations at the sub-nanometer scale is homogeneous. Near-stoichiometric films were found to be single crystal, with planar surface morphology, whereas the Zn-rich or Zn-poor films were polycrystalline, with surfaces consisting of crystallites and facets. Room temperature Raman spectra support the presence of significant cation disorder in the crystal. The measured CL and PL luminescence peaks show PL from the $ZnGeN_2$ layer reminiscent of the "yellow band" PL commonly observed in GaN.

Chapter 6

Experimental determination of the valence band offsets of $ZnGeN_2$ and

$(ZnGe)_{0.94}Ga_{0.12}N_2$ with GaN

6.1 Introduction

To date only calculations of the valence band offset (VBO) between the $Pna2_1$ phase of pure $ZnGeN_2$ and GaN have been reported. Punya et al. [43, 44] used DFT employing the local density approximation (LDA) for the electrostatic potential profile, and quasiparticle self-consistent GW calculations (where G is the one-electron Green's function and W the screened Coulomb interaction) for the band edge positions relative to the average electrostatic potential in each material. They included the strain effects in $ZnGeN_2$ to match the in-plane lattice constant to the unstrained GaN substrate. They found that the valence band maximum (VBM) of $ZnGeN_2$ lies well above that of GaN [43, 44]: the predicted VBO at the $ZnGeN_2$/GaN heterointerface with the normals along the $Pbn2_1$ [100], [010] and [001] directions were reported to be 1.44 eV, 1.36 eV, and 1.38 eV, respectively [44]. On the other hand, the "natural" valence band offset was determined from the calculated electron affinity values to be 0.49 eV for the (100) orientation of $Pbn2_1$ $ZnGeN_2$ [79]. The effect of the Zn $3d$ bands on the VBM can explain the positive VBO, with the valence band of $ZnGeN_2$ above that of GaN [43, 110]. The Zn $3d$ bands have stronger hybridization than the Ga $3d$ bands since they lie significantly closer to the VBM as compared to the Ga $3d$ bands. This situation will cause the VBM in $ZnGeN_2$ to move to a higher energy as compared to that in GaN [43, 110].

Recently, a DFT-based calculation using a hybrid functional [111] for surfaces combined with the electron-affinity rule found the natural VBM of $ZnGeN_2$ to be 0.28 eV below that of GaN, a result that is strikingly different from the results of Ref. [43] and [44]. The origin of this discrepancy is presently not clear.

In this chapter, we present the measurements of the valence band offsets of $(ZnGe)_{1-x}Ga_{2x}N_2$ with GaN for $x = 0$ ($ZnGeN_2$) and $x = 0.06$ ($(ZnGe)_{0.94}Ga_{0.12}N_2$) at the respective heterointerfaces using X-ray photoemission spectroscopy (XPS).

6.2 Experimental details

6.2.1 Sample growths

For this study, two $ZnGeN_2$ and two $(ZnGe)_{1-x}Ga_{2x}N_2$ samples were grown on commercially purchased GaN/c-sapphire templates in R2 reactor of the nitride MOCVD system at the Nanotech West Lab of the Ohio State University. DEZn, GeH_4 and NH_3 were used as the precursors for Zn, Ge, and N, respectively. Nitrogen (N_2) was used as the carrier gas. The total reactor pressure was set at 500 Torr. The DEZn/GeH_4 molar flow rate was optimized for obtaining single crystalline thin films with stoichiometric cation composition (Zn/(Zn+Ge) = 0.5) and smooth surface morphology. The detailed growth technique was described in section 5.2 of this book and has been reported in Ref. [60]. Two $ZnGeN_2$ samples were grown on GaN templates at growth temperature $T_G = 650$ °C, for 10 s and

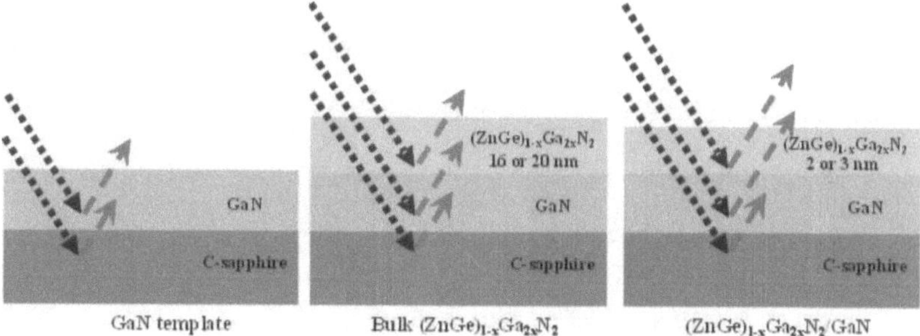

Figure 6.1 Schematics of the bulk GaN, bulk $(ZnGe)_{1-x}Ga_{2x}N_2$ and $(ZnGe)_{1-x}Ga_{2x}N_2$/GaN heterostructure samples used to determine the valence band offset of $ZnGeN_2$/GaN and $(ZnGe)_{0.94}Ga_{0.12}N_2$/GaN using XPS. The thicknesses of the bulk $ZnGeN_2$ and $(ZnGe)_{0.94}Ga_{0.12}N_2$ samples are 16 nm and 20 nm, respectively. The thicknesses of the $ZnGeN_2$ and $(ZnGe)_{0.94}Ga_{0.12}N_2$ layers in the heterostructure samples are 2 nm and 3 nm, respectively. The blue dotted lines indicate the incident X-ray beam and red dashed lines indicate the emitted photoelectrons. The arrows indicate the directions of propagation. Short red arrows indicate that the photoelectrons are absorbed before escaping from the sample. [64]

80 s, using the same growth conditions for both. Two $(ZnGe)_{1-x}Ga_{2x}N_2$ samples were grown at T_G = 735 °C for 23 s and 165 s on GaN templates. An in-situ reflectometer was used to estimate the thicknesses of the films. Growth rates were also confirmed from cross-sectional SEM images of thicker films. The thicknesses of the $ZnGeN_2$ samples were ~2 nm and ~16 nm for the growth durations of 10 s and 80 s, respectively whereas the thicknesses of the $(ZnGe)_{1-x}Ga_{2x}N_2$ samples were ~3 nm and ~20 nm for the growth

84

durations of 23 s and 165 s, respectively. It is shown later that, for XPS measurements, the 16-nm-thick $ZnGeN_2$ and the 20-nm-thick $(ZnGe)_{1-x}Ga_{2x}N_2$ films work as bulk $ZnGeN_2$ and bulk $(ZnGe)_{1-x}Ga_{2x}N_2$, respectively, whereas the 2-nm-thick $ZnGeN_2$ and the 3-nm-thick $(ZnGe)_{1-x}Ga_{2x}N_2$ films work as $ZnGeN_2/GaN$ and $(ZnGe)_{1-x}Ga_{2x}N_2/GaN$ heterostructures, respectively.

6.2.2 Material characterizations

The quality of the heterointerfaces was investigated using high magnification STEM images captured by a Thermofisher probe-corrected Titan STEM operated at 300 kV. SEM imaging were carried out by a Helios Nanolab 600 and an FEI Apreo LoVac analytical SEM. Surface roughness of the films were determined from AFM images using a Bruker Icon 3 AFM. XRD measurements were carried out using a Bruker D8 Discover XRD with the Cu $K\alpha$ source. The XPS measurements on the $ZnGeN_2$ samples were carried out by Brent A Noesges from Prof. Leonard J. Brillson's group using a PHI 5000 VersaProbeTM system equipped with a Scanning XPS Microprobe X-ray source with hυ (Al $K\alpha$) = 1486.6 eV, FWHM$\leq$ 0.5 eV. The XPS measurements on the $(ZnGe)_{1-x}Ga_{2x}N_2$ samples were performed using a Kratos Axis Ultra X-ray photoelectron spectrometer using a monochromatic Al $K\alpha$ X-ray source, E_{photon} = 1486.6 eV. The measurements were performed on as-grown samples without performing additional surface cleaning before or after loading into the XPS chambers to prevent potential surface contamination or structural damage [25]. The samples were stored in the chamber for 18 hours under high vacuum prior to the XPS measurements.

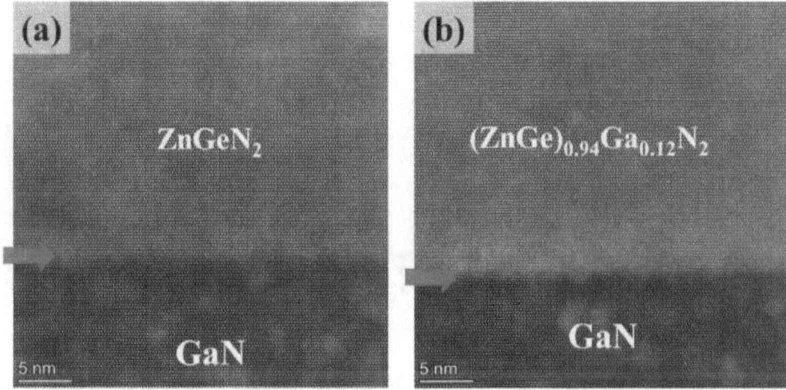

Figure 6.2 High magnification STEM images showing the interfaces of (a) a ZnGeN$_2$ film and (b) a (ZnGe)$_{1-x}$Ga$_{2x}$N$_2$ film on GaN. The interface is marked by the red arrows. [64]

6.3 Crystalline, and morphological properties of the samples

Figure 6.1 shows schematic diagrams of the samples. Figures 6.2(a) and 6.2(b) present the high magnification STEM images of a ZnGeN$_2$ and a (ZnGe)$_{0.94}$Ga$_{0.12}$N$_2$ film grown on GaN grown under identical conditions to the corresponding samples used to measure the band offsets in this work. The red arrows mark the interfaces which show clear contrasts between the substrate and the films. The STEM images demonstrate the high-quality interfaces between the binary GaN and ternary ZnGeN$_2$ or quaternary (ZnGe)$_{1-x}$Ga$_{2x}$N$_2$ lattices. To investigate the surface morphologies, FESEM images and AFM images were obtained from a ~350 nm thick ZnGeN$_2$ film and a ~1 μm thick (ZnGe)$_{1-x}$Ga$_{2x}$N$_2$ film, which were grown using the identical growth conditions as those described above. Plan-view SEM images in Fig. 6.3 (a), and 6.3 (b) show planar surfaces for both

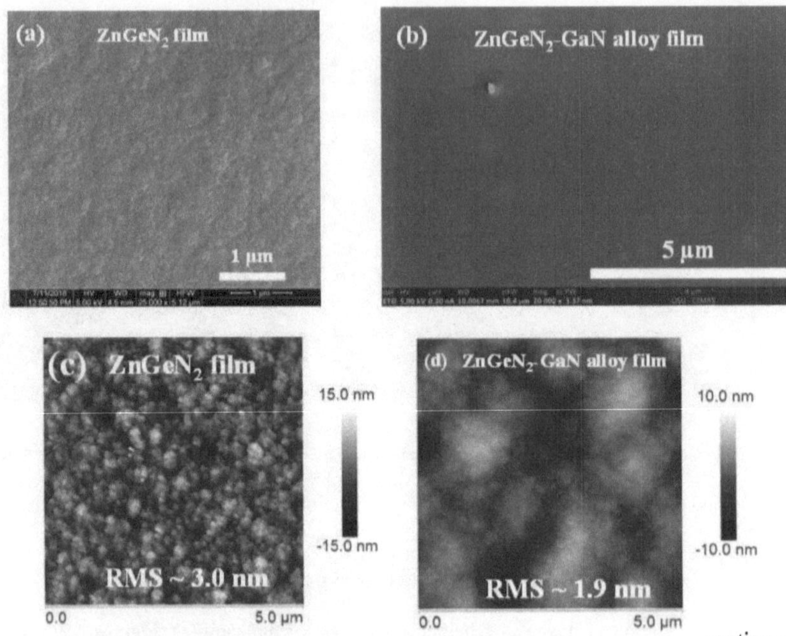

continued

Figure 6.3 (a, b) Plan-view FESEM images and (c, d) 5 µm × 5 µm AFM images of a ~350 nm thick $ZnGeN_2$ film (a, c) and a ~1 µm thick $(ZnGe)_{1-x}Ga_{2x}N_2$ film (b, d) grown using identical growth conditions as those used for determining the valence band offset in this work. (e) XRD 2θ-ω scan profile obtained from a representative $ZnGeN_2$ film with stoichiometric cation composition grown on GaN/c-sapphire template for a 2θ range from 30° to 90°. The inset shows the zoomed in view around the $ZnGeN_2$ (002) and (004) peaks to clearly show the separation with respective GaN peaks. (f) ω-rocking curve around the $ZnGeN_2$ (002) peak obtained from a single crystalline $ZnGeN_2$ film grown on c-sapphire. [64]

Figure 6.3 continued

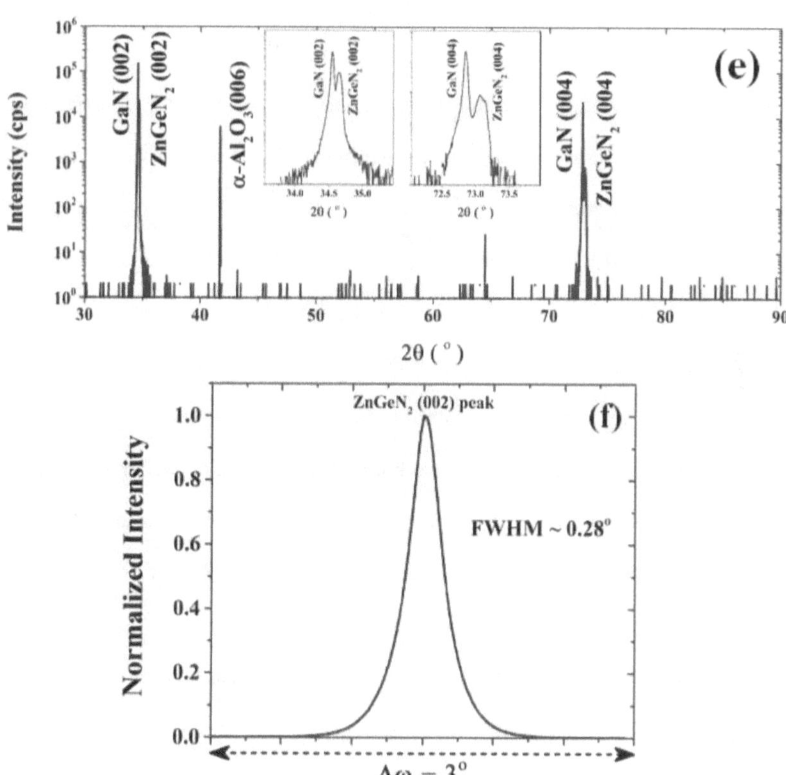

films. The RMS roughness values obtained from 5 μm × 5 μm AFM images (Figs. 6.3(c), and 6.3(d)) were 3.0 nm and 1.9 nm, respectively. Figure 6.4(e) shows the XRD 2θ-ω scan profile obtained from a representative ~ 1 μm thick $ZnGeN_2$ film with stoichiometric cation composition grown on GaN/c-sapphire template. Within the scanned range ($2\theta = 30$ - $90°$), only five peaks associated with GaN (002), $ZnGeN_2$ (002), α-Al_2O_3 (006), GaN (004), and $ZnGeN_2$ (004) planes, respectively, were observed. The inset shows zoomed in view near the (002) and (004) peaks of GaN and $ZnGeN_2$. No peaks corresponding to secondary phases of $ZnGeN_2$, for example, Zn_3N_2 or Ge_3N_4 were observed. Based on the XRD 2θ-ω scan profiles, $ZnGeN_2$ and $(ZnGe)_{1-x}Ga_{2x}N_2$ thin films grown under various similar conditions to those used in this work were single crystalline and phase pure [60, 61]. The absence of secondary phases has also been confirmed by atom probe tomography measurements which has already been reported in Ref. [60]. For $ZnGeN_2$ films grown on GaN, very close XRD 2θ positions of the film and the substrate (e.g., for ((002) peaks $\Delta 2\theta$ ~ 0.15°) [60]) makes it challenging to separate the signals between the two from XRD ω-rocking curves. For $ZnGeN_2$ films grown on c-sapphire substrates, a *full width at half maxima* (FWHM) of the ω-rocking curve around the (002) peak as low as 0.28° was obtained (Fig. 6.3(f)). Crystalline quality of the $ZnGeN_2$ films grown on GaN (lattice mismatch < 1%) is significantly improved as compared to the $ZnGeN_2$ films grown on c-sapphire (~16% lattice mismatch). A detailed investigation of crystal structural and surface morphological properties of $ZnGeN_2$ films was reported in Ref. [60].

6.4 Kraut's method for determination of band offset at heterointerface

The valence band offset, ΔE_V, between two materials A and B can be determined by using Kraut's method [112] according to which

$$\Delta E_v = \left(E_{CL,b}^B - E_v^B\right) - \left(E_{CL,b}^A - E_v^A\right) - \left(E_{CL,i}^B - E_{CL,i}^A\right) \tag{6.1}$$

where, $E_{CL,b}^{A/B}$ is the binding energy of the core level in the bulk material A/B, $E_{CL,i}^{A/B}$ is the binding energy of the core level of the material A/B determined from the heterostructure between A and B and $E_V^{A/B}$ is the position of the VBM in the bulk material A/B. The positions of the core levels are determined with respect to the Fermi level. A positive value of ΔE_v would indicate the VBM of B to be above that of A.

XPS is a commonly used technique to determine the parameters in Eq. (6.1) [113, 114]. The parameters in the first two terms in Eq. (6.1) are determined using XPS measurements of the bulk samples of materials A and B. The parameters in the third term are determined using XPS measurement of a heterostructure formed between the two materials. In order to be able to measure the core level of the bottom layer of the two-layer heterostructure using XPS, the photoelectrons from this layer need to be detected. A thick top layer can absorb all of the photoelectrons emitted from the bottom layer due to the finite escape depth of the photoelectrons [113, 115], as illustrated in Fig. 6.1. Therefore, the top layer needs to be sufficiently thin so that a reasonable number of photoelectrons from the bottom layer can escape and be detected.

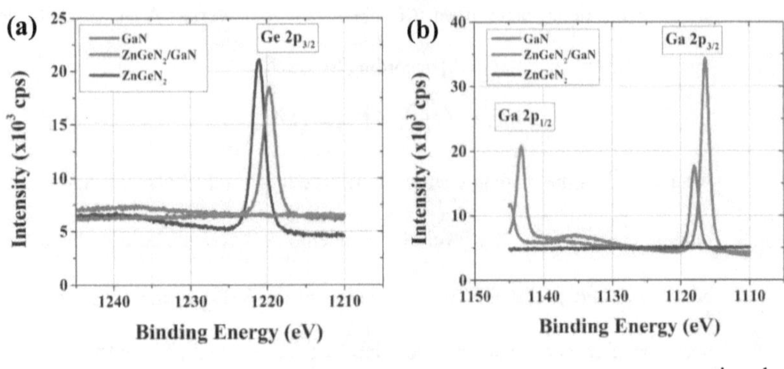

continued

Figure 6.4 The XPS spectra near the (a) Ge $2p$ region, (b) *Ga* $2p$ region and (c) Zn $2p$ region obtained from the GaN, 16 nm ZnGeN₂ and the 2 nm ZnGeN₂/GaN heterostructure samples. (d-f) XPS spectra near the Ge $3d$ (left panels), Ga $3d$ (center panels), and Zn $3d$ (right panels) bands obtained from the GaN, 16 nm ZnGeN₂ and the 2 nm ZnGeN₂/GaN heterostructure samples, respectively. The original spectra are shown by the black circles. The dashed blue curves are the fitted peaks obtained assuming a Voigt shape and Shirley background. The background is shown by the green dash-dot lines. The Zn $3d$ peaks overlapped with the N $2p$ peaks which are marked in respective panels. The full width at half maxima of the fitted peaks 1.7 eV, 1.0 eV and 1.6-1.7 eV for Ge $3d$, Ga $3d$ and Zn $3d$ bands, respectively. (g) The XPS spectra showing the valence band edges of the GaN (magenta circles) and the 16 nm ZnGeN₂ (blue squares) samples. [64]

Figure 6.4 continued

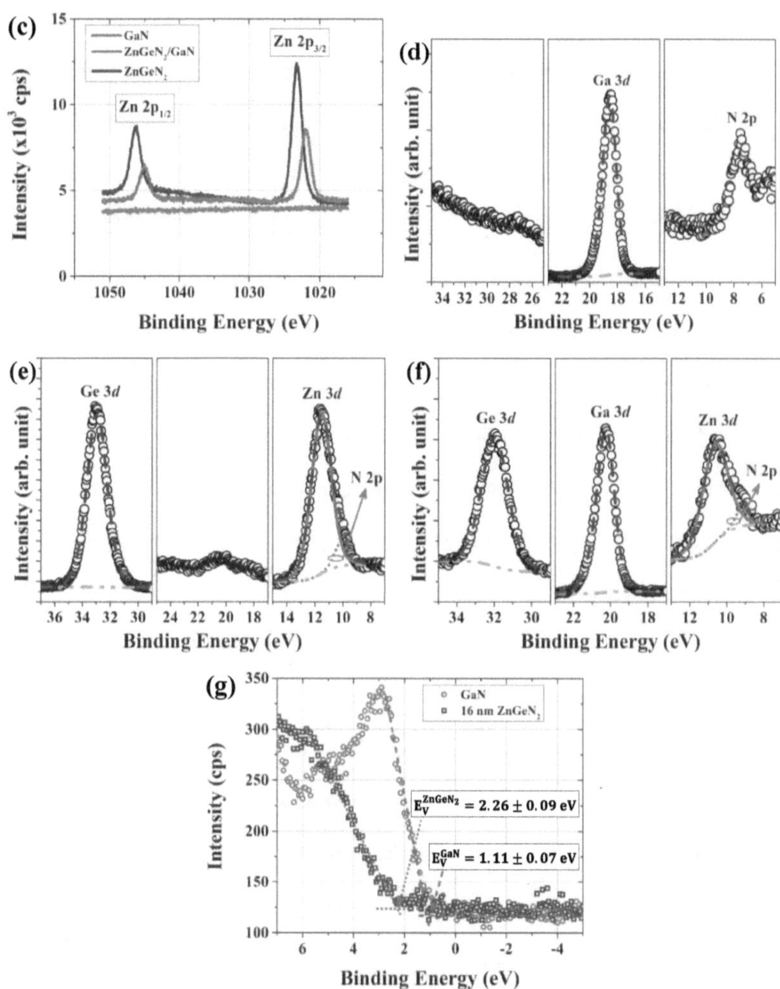

6.5 Valence band offset between ZnGeN₂ and GaN

With a view to determining the band offset between $ZnGeN_2$ and GaN at a $ZnGeN_2$/GaN heterointerface, XPS measurements were performed on the GaN template, the 16-nm-thick $ZnGeN_2$ layer and the 2-nm-thick $ZnGeN_2$/GaN heterostructure samples. The binding energies of the adventitious C $1s$ peaks in these three samples were measured to be 283.9 eV, 286.1 eV, and 284.9 eV, respectively. Figures 6.4 (a-c) present the XPS spectra near the Ge $2p$, Ga $2p$ and Zn $2p$ core-level positions, respectively, for all three samples. As can be seen here, the Ga $2p$ peak is absent in the spectra from the 16-nm-thick $ZnGeN_2$ sample, which indicates that this sample has sufficient thickness to be characteristic of bulk $ZnGeN_2$. In addition, only the Ga $2p$ peak was observed in the XPS spectra of the GaN template. The XPS spectra of the 2-nm-thick $ZnGeN_2$/GaN sample showed Ge $2p$ and Zn $2p$ as well as Ga $2p$ peaks. Therefore, this sample is characteristic of $ZnGeN_2$/GaN heterostructures.

The XPS spectra near the Ge $3d$, Ga $3d$, and Zn $3d$ bands obtained from the GaN, 16-nm-thick $ZnGeN_2$ and 2-nm-thick $ZnGeN_2$/GaN heterostructure samples are shown in Figs. 6.4(d), 6.4(e), and 6.4(f), respectively, by the black circles. The blue dashed lines were fitted to the peaks by assuming a Voigt line shape and Shirley background. The Shirley backgrounds are shown by the green dash-dot curves. The Zn $3d$ peaks overlapped with the N $2p$ peaks for both the 16-nm-thick $ZnGeN_2$ and the 2-nm-thick $ZnGeN_2$/GaN heterostructure samples, shown in in Figs. 6.4(e), and 6.4(f), respectively. For the GaN sample (magenta spectra), only the Ga $3d$ peak, at a binding energy of 18.6 eV, was observed. On the other hand, for the $ZnGeN_2$ sample (blue spectra), only the Ge $3d$ and Zn

Table 6.1 Positions of the bulk valence band maxima and the core Ga, Zn and Ge levels extracted from the XPS spectra, from the GaN template, the 16 nm $ZnGeN_2$ (bulk) and the 2 nm $ZnGeN_2$/GaN (heterostructure) samples. The binding energies of the core levels were determined by peak fitting assuming a Voigt shape and Shirley background. No correction in the peak positions was done. The binding energies of adventitious C 1s peaks are listed for reference.

Sample	C	Ga		Zn		Ge		Ev
	E_{1s} (eV)	$E_{Ga-2p3/2}$ (eV)	E_{Ga-3d} (eV)	$E_{Zn-2p3/2}$ (eV)	E_{Zn-3d} (eV)	$E_{Ge-2p3/2}$ (eV)	E_{Ge-3d} (eV)	VBM (eV)
GaN	283.9	1116.4	18.6	-	-	-	-	1.1 ±0.07
$ZnGeN_2$	286.1	-	-	1023.1	11.4	1221.0	33.0	2.26±0.09
$ZnGeN_2$/GaN	284.9	1118.0	20.3	1022.0	10.5	1219.7	31.9	-

$3d$ peaks were observed, at binding energies 33.0 eV and 11.4 eV, respectively. All three (Ge, Ga, and Zn) $3d$ peaks were observed in the spectra from the 2 nm $ZnGeN_2$/GaN heterostructure sample (red spectra), at binding energies 31.9 eV, 20.25 eV and 10.5 eV, respectively. The FWHM values were 1.0 eV for Ga $3d$ peaks in Fig. 4(d) and Fig. 4(f), 1.7 eV for Ge $3d$ peaks and 1.6-1.7 eV for Zn $3d$ peaks. Please note that no corrections in the peak positions were done since only the differences between the spectral features from any given sample are used to calculate the VBO, namely, core level differences between GaN substrate and $ZnGeN_2$ overlayer on the one hand and VBM vs. the same core levels from separate bulk-like samples (GaN substrate and thick $ZnGeN_2$ sample) on the other hand. Any offset in absolute binding energies between different samples in Eq. 6.1 will be cancelled out. This is precisely the advantage of the Kraut method. The measured binding energies for the core levels are within the range of corresponding values reported previously [114].

As noted earlier, the binding energies of the Zn and Ge core levels obtained from the ZnGeN$_2$ sample give $E_{CL,b}^{ZnGeN_2}$, the binding energies of the Ga core levels obtained from the GaN sample give $E_{CL,b}^{GaN}$. The corresponding values from the 2-nm-thick ZnGeN$_2$/GaN heterostructure sample give $E_{CL,i}^{ZnGeN_2}$ or $E_{CL,i}^{GaN}$. The VBM values, E_V^{GaN} and $E_V^{ZnGeN_2}$, were obtained from the GaN and 16-nm-thick ZnGeN$_2$ samples, respectively. Figure 6.4(g) presents the XPS spectra near the valence band edge of the 16-nm-thick ZnGeN$_2$ sample and the GaN template. The VBM values were calculated from the intersection of the straight line fitted over the leading edge of the XPS spectra with the average of the flat portion of the spectra for energies above the valence band maxima. Table 6.1 lists the values of the VBM as well as the core level binding energies. The calculated VBO between ZnGeN$_2$ and GaN from Eq. (6.1) is 1.45±0.15 eV when Zn $3d$ and Ga $3d$ core-level energies are used and 1.65±0.15 eV when Ge $3d$ and Ga $3d$ core-level energies are used with the average being 1.55±0.15 eV. The calculated VBO value using Zn $2p_{3/2}$ and Ga $2p_{3/2}$, and Ge $2p_{3/2}$ and Ga $2p_{3/2}$ core level energies is 1.54±0.15 eV, and 1.74±0.15 eV, respectively, which are consistent with the values calculated using corresponding $3d$ core-level energies. These VBO values are in reasonable agreement with those obtained from explicit interface calculations (1.4 eV) [43, 44] rather than the electron-affinity based values [79, 111].

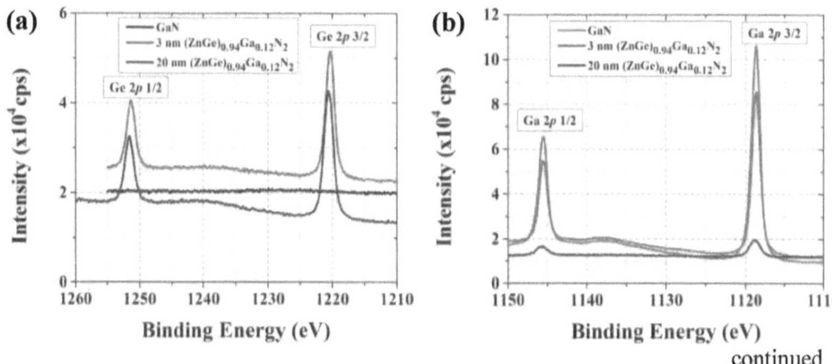

continued

Figure 6.5 The XPS spectra near the (a) Ge $2p$ region, (b) *Ga* $2p$ region and (c) Zn $2p$ region obtained from the GaN, 20 nm $(ZnGe)_{0.94}Ga_{0.12}N_2$ and the 3 nm $(ZnGe)_{0.94}Ga_{0.12}N_2$ /GaN heterostructure samples. (d-f) XPS spectra near the Ge $3d$ (left panels), Ga $3d$ (center panels), and Zn $3d$ (right panels) bands obtained from the GaN, 20 nm $(ZnGe)_{0.94}Ga_{0.12}N_2$ and the 3 nm $(ZnGe)_{0.94}Ga_{0.12}N_2$/GaN heterostructure samples, respectively. The original spectra are shown by the black circles. The dashed blue curves are the fitted peaks obtained assuming a Voigt shape and Shirley background. The background is shown by the green dash-dot lines. The Zn $3d$ peaks overlapped with the N $2p$ peaks which are marked in respective panels. For the 20 nm $(ZnGe)_{0.94}Ga_{0.12}N_2$ sample, two components were required to fit the Ge $3d$ peak. The full width at half maxima of the fitted peaks 1.4 eV, 1.0 eV and 1.4-1.5 eV for Ge $3d$, Ga $3d$ and Zn $3d$ bands, respectively. (g) The XPS spectra showing the valence band edges of the GaN (magenta circles) and the 20 nm $(ZnGe)_{0.94}Ga_{0.12}N_2$ (blue squares) sample. [64]

Figure 6.5 continued

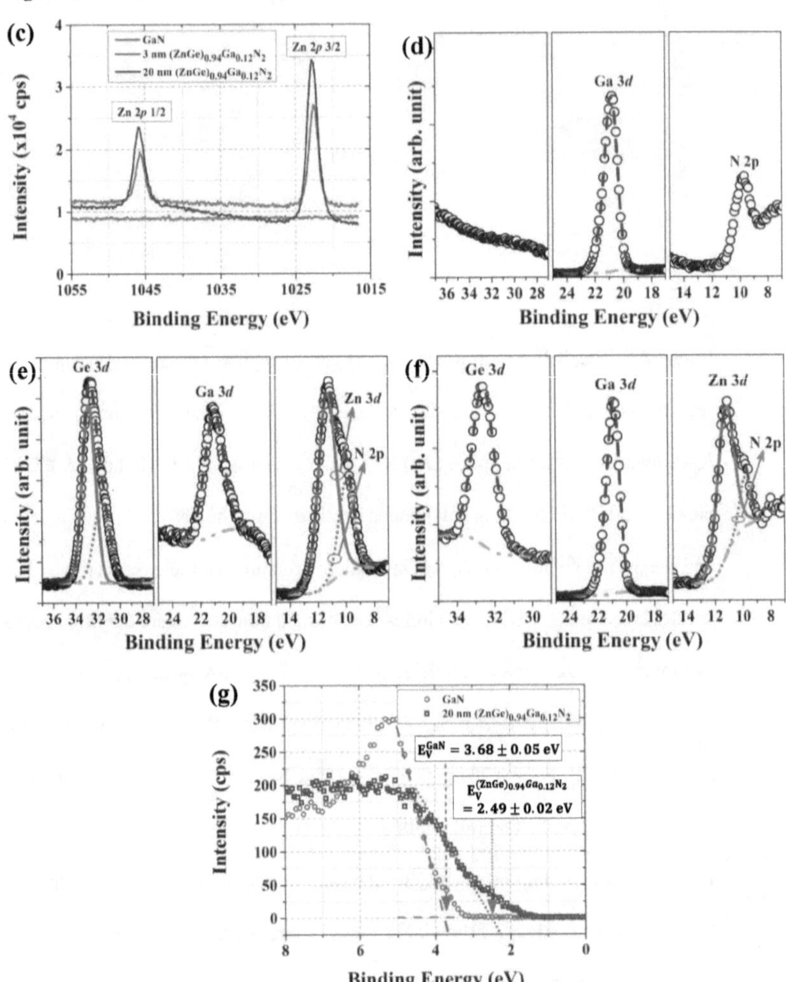

6.6 Valence band offset between $(ZnGe)_{0.94}Ga_{0.12}N_2$ and GaN

The same approach was used to determine the VBO of $(ZnGe)_{0.94}Ga_{0.12}N_2$ with GaN. The XPS spectra collected from $(ZnGe)_{0.94}Ga_{0.12}N_2$ samples are shown in Fig. 6.5. The spectra measured from a bare piece of GaN template on which the $(ZnGe)_{0.94}Ga_{0.12}N_2$ samples were grown are also shown. The binding energies of the adventitious C $1s$ peaks in these three samples were measured to be 286.1 eV, 285.8 eV, and 285.6 eV, respectively. The Ge $2p$ and Zn $2p$ peaks were observed in Fig. 6.5(a) and Fig. 6.5(c), respectively, from both the 3 nm and 20 nm thick $(ZnGe)_{0.94}Ga_{0.12}N_2$ samples, but not from the GaN template. The Ga $2p$ peak was observed in all three samples. The 20-nm-thick $(ZnGe)_{0.94}Ga_{0.12}N_2$ is sufficiently thick to absorb all of the XPS signals originating in the underlying GaN substrate. Therefore, the Ga $2p$ peaks in the spectra from the 20-nm-thick $(ZnGe)_{0.94}Ga_{0.12}N_2$ sample, shown in blue, can be assigned to the Ga atoms in the top film. Figures 5(d-f) show the XPS spectra near the Ge $3d$, Ga $3d$, and Zn $3d$ bands obtained from the GaN, 20-nm-thick $(ZnGe)_{0.94}Ga_{0.12}N_2$ and 3-nm-thick $(ZnGe)_{0.94}Ga_{0.12}N_2$/GaN heterostructure samples, respectively, in black circles. The dashed blue curves in figures 5(d-f) are the fitted peaks assuming Voigt peak shapes and Shirley background. The overlapped Zn $3d$ and N $2p$ peaks are marked in figures 6.5(e) and 6.5(f). In the case of the 20-nm-thick $(ZnGe)_{0.94}Ga_{0.12}N_2$ sample in figure 6.5(e) two components were necessary to fit the Ge $3d$ peak – The peak at the lower binding energy probably corresponds to a lower charge state of Ge present in the sample [116]. The binding energies of the Ge $3d$ and Zn $3d$ peaks were determined to be 32.7 and 11.3 eV, respectively, for the 20-nm-thick $(ZnGe)_{0.94}Ga_{0.12}N_2$ sample, and 32.6 eV and 11.2 eV, respectively, for the 3-nm-thick

Table 6.2 Position of the bulk valence band maxima and the core Ga, Zn and Ge levels extracted from the XPS spectra of the GaN template, the 20 nm $(ZnGe)_{0.94}Ga_{0.12}N_2$ (bulk) and the 3 nm $(ZnGe)_{0.94}Ga_{0.12}N_2$/GaN (heterostructure) samples. The binding energies of the core levels were determined by peak fitting assuming a Voigt shape and Shirley background. No correction in the peak positions was done. The binding energies of adventitious C 1s peaks are listed for reference.

Sample	C	Ga		Zn		Ge		E_V
	E_{1s} (eV)	$E_{Ga\text{-}2p3/2}$ (eV)	$E_{Ga\text{-}3d}$ (eV)	$E_{Zn\text{-}2p3/2}$ (eV)	$E_{Zn\text{-}3d}$ (eV)	$E_{Ge\text{-}2p3/2}$ (eV)	$E_{Ge\text{-}3d}$ (eV)	VBM (eV)
GaN	286.1	1118.7	20.9	-	-	-	-	3.68±0.05
$(ZnGe)_{0.94}Ga_{0.12}N_2$	285.8	1118.8	21.0	1022.7	11.3	1220.5	32.7	2.49±0.02
$(ZnGe)_{0.94}Ga_{0.12}N_2$/GaN	285.6	1118.6	20.9	1022.6	11.2	1220.3	32.6	-

$(ZnGe)_{0.94}Ga_{0.12}N_2$/GaN heterostructure sample. The FWHM values were 1.4 eV for Ge $3d$ peaks and 1.4-1.6 eV for Zn $3d$ peaks. The binding energy of the Ga $3d$ peak was 20.9 eV in both the bulk GaN and the 3-nm-thick $(ZnGe)_{0.94}Ga_{0.12}N_2$/GaN heterostructure sample. The FWHM of these Ga $3d$ peaks are 1.0 eV. The positions of the valence band maxima in the 20-nm-thick $(ZnGe)_{0.94}Ga_{0.12}N_2$ (bulk) sample and the GaN template are marked in Fig. 6.5(g). These are at 2.49±0.02 eV and 3.68±0.05 eV, respectively. The difference (2.58 eV) between the positions of the VBM of GaN templates in Fig. 6.4(g) and Fig. 6.5(g) is comparable to the difference (2.2 eV) between the binding energies of the adventitious C 1s levels of the two samples, which is probably caused by the difference in electrical conductivity of the two templates. The difference in electrical conductivity can result from different concentrations of unintentional impurities in the film, for example, C, which in turn can cause shifts in the core-level positions [117].

The core energy levels and the positions of the VBM determined for the $(ZnGe)_{0.94}Ga_{0.12}N_2$ samples are listed in Table 6.2. The VBO between $(ZnGe)_{0.94}Ga_{0.12}N_2$ and GaN calculated using either Zn or Ge $3d$ core levels is 1.29 ± 0.2 eV, which is 0.26 eV lower than the average VBO determined for $ZnGeN_2$ using the $3d$ core-level energies. This lowering of the VBO is attributed to the reduced effect of the Zn $3d$ bands in the $(ZnGe)_{0.94}Ga_{0.12}N_2$ compared to in $ZnGeN_2$. Assuming a linear dependence of the VBO with composition x of $(ZnGe)_{1-x}Ga_{2x}N_2$, the predicted VBO of $(ZnGe)_{0.94}Ga_{0.12}N_2$ from theoretically calculated VBO of $ZnGeN_2$ [44] would be 1.32 eV, which is very close to the experimentally determined value. The error bars in the determined VBO values are mainly due to the uncertainties in the determined VBM values. The calculated total polarization difference at the $ZnGeN_2$/GaN heterointerface is very small [68, 111], therefore, errors in the determined VBO values due to polarization induced band bending is expected to be small. For instance, the $ZnGeN_2$/GaN heterostructure VBOs, predicted by Jaroenjittichai et al. [44], along the polar and the non-polar directions differ by < 0.1 eV.

6.7 Band alignment of $ZnGeN_2$, and $(ZnGe)_{0.94}Ga_{0.12}N_2$ with GaN

The conduction band offsets were calculated using the determined VBO values and the band gap of the materials using the relation $\Delta E_C = \Delta E_V + (E_g^{(ZnGe)_{1-x}(Ga)_{2x}N_2} - E_g^{GaN})$. The band gaps of $ZnGeN_2$ and $(ZnGe)_{0.94}Ga_{0.12}N_2$ were estimated by analyzing the inelastic energy loss features of N 1s peaks in the XPS spectra of 16-nm-thick $ZnGeN_2$ and 20-nm-thick $(ZnGe)_{0.94}Ga_{0.12}N_2$ samples. Before reaching the XPS detector, a photo-emitted electron can lose energy to (i) high frequency plasma oscillations in the valence band, (ii) surface plasmons and (iii) another electron transitioning from the valence band into the conduction band. The band-to-band transition involves the minimum-energy process among the three, and therefore, the band gap of the material determines the lower limit of the energy loss of the photo-emitted electrons due to inelastic processes [118]. Energy loss features of the N $1s$ peak in XPS spectra were previously used to determine the bandgap of Si_3N_4 [119]. Figures 6.6(a) and 6.6(b) show the XPS spectra near the N $1s$ core levels obtained from the 16-nm-thick $ZnGeN_2$ and 20-nm-thick $(ZnGe)_{0.94}Ga_{0.12}N_2$ samples, respectively. The approximate onsets of the inelastic energy losses were estimated by linear fitting to the energy-loss features on the higher binding energy side of the N $1s$ peak. The difference between the binding energies corresponding to the N $1s$ peak and the onset of the inelastic energy loss provides the bandgap of the epilayer. The band gaps found by this method are 3.0±0.2 eV for $ZnGeN_2$ and 3.1±0.2 eV for $(ZnGe)_{0.94}Ga_{0.12}N_2$. The error bars correspond to the standard deviation of the estimated onset of the inelastic energy losses. The slight increase in the band gap of the 50:50 alloy of $ZnGeN_2$-GaN alloy $((ZnGe)_{0.5}GaN_2)$ as compared to those of $ZnGeN_2$ or GaN was predicted by first-principles calculations [110]. The lower values of the band gaps as compared to the predicted values are probably due to the presence of disorder in the cation sublattice [26]. The band alignments of $ZnGeN_2$ and $(ZnGe)_{0.94}Ga_{0.12}N_2$ with GaN are shown in Fig. 6.6(c) using the average of the VBO values determined using the Zn $3d$ and Ge $3d$ bands.

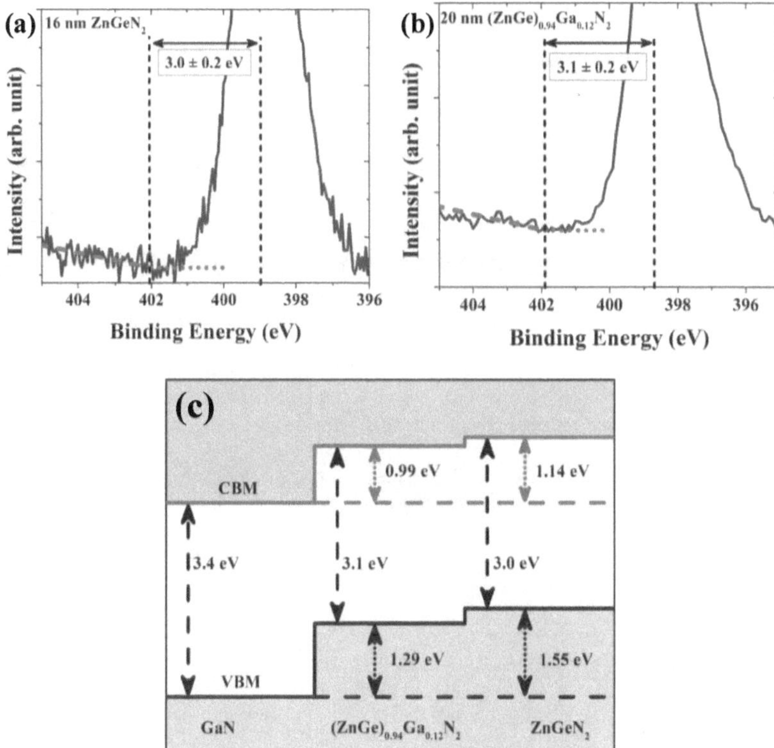

Figure 6.6 The XPS spectra showing the N $1s$ peak and onset of inelastic energy loss obtained from (a) 16-nm-thick $ZnGeN_2$ and (b) 20 nm $(ZnGe)_{0.94}Ga_{0.12}N_2$ samples. (c) Band alignments of $(ZnGe)_{1-x}Ga_{2x}N_2$ (x = 0, and 0.06) with GaN. The average of the determined valence band offsets using the Zn $3d$ and Ge $3d$ bands were used to align the valence band maxima of $(ZnGe)_{0.94}Ga_{0.12}N_2$ and $ZnGeN_2$ relative to the VBM of GaN. [64]

6.8 Conclusions

In conclusion, the valence band offsets of MOCVD-grown $(ZnGe)_{1-x}Ga_{2x}N_2$ with GaN, with $x = 0$ and 0.06, measured using XPS, are presented for the first time to the best of our knowledge. The measured VBO for $ZnGeN_2$ (1.45 eV - 1.65 eV) are comparable to the predicted value from first-principles calculations using explicit interface calculations [43, 44]. For $(ZnGe)_{0.94}Ga_{0.12}N_2$, the VBO was measured to be 1.29 eV, which is very close to the predicted value from theoretically calculated VBO of $ZnGeN_2$ assuming a linear dependence of VBO on composition. The results from this study will expand device designs based on pure III-nitrides to III-nitrides/II-IV-N_2, which can potentially address key challenges in III-nitride based electronic and optoelectronic device technologies.

Chapter 7

MOCVD growth and characterization of GaN/InGaN/GaN, and GaN/InGaN/ZnGeN$_2$/InGaN/GaN quantum wells

7.1 Introduction

Prerequisites for successful implementation of InGaN/ZnGeN$_2$/InGaN heterostructure QWs are (i) optimization of the GaN/InGaN/GaN QWs structure and (ii) development and optimization of growth conditions for ZnGeN$_2$. The latter was discussed in Chapter 4 and Chapter 5 of this book. In this chapter, we present, first, the work on calibration of conventional GaN/InGaN/GaN QWs growth in our reactor and then, the work on implementation of GaN/InGaN/ZnGeN$_2$/InGaN/GaN heterostructure QWs. The structure of the grown QWs was investigated using STEM imaging and XRD measurements. The emission properties were investigated using CL and PL spectroscopy. The luminescence spectra of GaN/InGaN/ZnGeN$_2$/InGaN/GaN QWs showed. one additional peak to those present in the reference conventional QW, which is at a lower energy position (red-shifted) than the band-to-band emission in the reference conventional GaN/InGaN/GaN QW.

104

7.2 Experimental details

7.2.1 MOCVD growth

The samples in this chapter were grown in reactor 2 (R2) of the nitride MOCVD system (CVD03) located at the Nanotech West laboratory. Commercially obtained GaN/c-sapphire templates were used as the substrates. Triethylgallium (TEGa), trimethylindium (TMIn), DEZn, GeH_4, and NH_3 were used as the precursors for Ga, In, Zn, Ge, and N, respectively. N_2 was used as the carrier gas. Prior to loading into the reactor, the templates were cleaned using acetone, and iso-propanol, rinsed with deionized water, and finally, blow dried using high purity N_2. Each growth was initiated by growing a ~175-200 nm GaN layer at 975 °C growth temperature and 200 Torr reactor pressure using trimethylgallium (TMGa) and NH_3 as the Ga and N precursors, respectively, and H_2 as the carrier and process gas. The temperature and pressure were then changed to those used for growth of QW structures. NH_3 was kept flowing during this period.

7.2.2 Characterizations

The emission properties of the QW samples were investigated using PL and CL spectroscopy. The PL measurements were carried out using a Horiba Fluorolog-3 system equipped with a single photon counter. A He-Cd laser or a Xe arc lamp was used as the excitation source. The laser wavelength was 325 nm. In case of the Xe lamp, a monochromator was used to select certain excitation wavelengths, for example, 325 nm, and 375 nm. CL measurements were performed using a Thermo Fisher Quattro Environmental Scanning Electron Microscope (ESEM) equipped with Horiba H-Clue CL detector at the Ohio State University Center for Electron Microscopy and Analysis.

The structural properties of the samples were investigated using XRD measurements as well as STEM imaging. The XRD measurements were performed using a Bruker Discover D8 XRD with Cu Kα (1.5406 Å wavelength) source. The STEM imaging was carried out at Ohio State University Center for Electron Microscopy and Analysis using a Thermofisher probe-corrected Titan STEM operated at 300 kV. EDS line scans were also performed using the same tool to determine the atomic fraction of the regarding elements.

7.3 Challenges to implementation of III-N/Zn-IV-N$_2$ heterostructure QW

7.3.1 Cross-contamination

One technical challenge for successful implementation of heterostructure devices is the possibility of cross-contamination of the layers by the constituent elements of the preceding layers. For example, unintentional incorporation of Ga during MOCVD growth of AlInN, up to the composition of as high as 45%, has been reported [120, 121]. The possible sources contributing to such unintentional Ga incorporation include the GaN templates used for epilayer growth [122], and the residual Ga in the growth chamber from previous growth-steps [120].

One of the aims of this work is to implement InGaN/ZnGeN$_2$ heterostructure based QW LEDs, which requires successive growths of III-N and II-IV-N$_2$ layers. We have noticed that the growth of III-N and Zn-IV-N$_2$ materials in the same MOCVD growth chamber results in cross-contamination in both III-Ns and Zn-IV-N$_2$s. The following results are evidence for cross-contamination of an InGaN film by Zn and Sn and a ZnGeN$_2$ film by Ga and In.

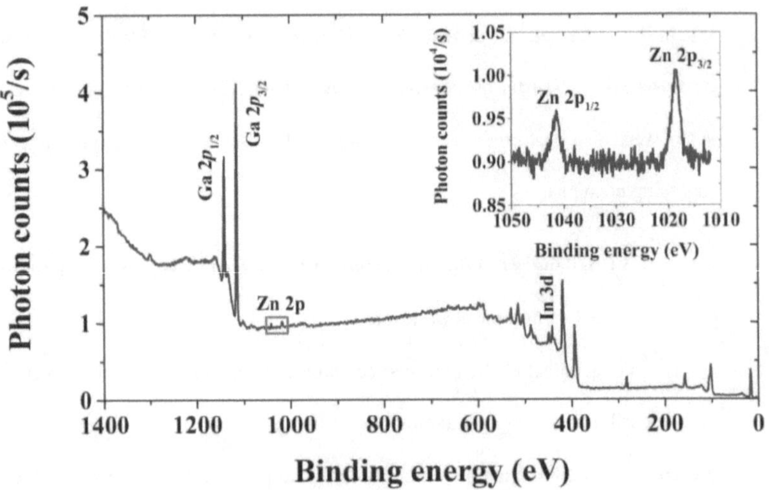

Figure 7.1 XPS survey spectra showing the unintentional Zn incorporation in MOCVD grown InGaN films on GaN. The inset shows the XPS spectra near the Zn 2p levels.

7.3.1.1 Unintentional Zn and Sn incorporation in InGaN film grown on a GaN

Figure 7.1 shows the XPS spectra of an InGaN films grown on GaN template using TEGa, TMIn and NH$_3$ as the precursors. N$_2$ was used as the carrier gas. No other gas was flowed into the chamber during the growth. The reactor was baked twice at 1020 °C for an average duration of 35 minutes per bake before the growth. The susceptor used for placing the substrate had never been used for ZnGeN$_2$ or ZnSnN$_2$ growth. The estimated thickness of the film is 60 nm from reflectivity profile.

In the XPS survey spectra in Fig. 7.1, Zn 2*p* and Sn 3*d* peaks are marked along with the Ga 2*p* and In 3*d* peaks. The XPS spectra near the Zn 2*p* and Sn 3*d* regions are shown in insets A and B, respectively. Zn and Sn are evidently present in the grown film, which were incorporated unintentionally even though no Zn or Sn precursors were flowed during the growth. The atomic composition of Ga, In and Zn determined from the survey spectra were 85±2%, 9±2%, and 2±0.5%, respectively.

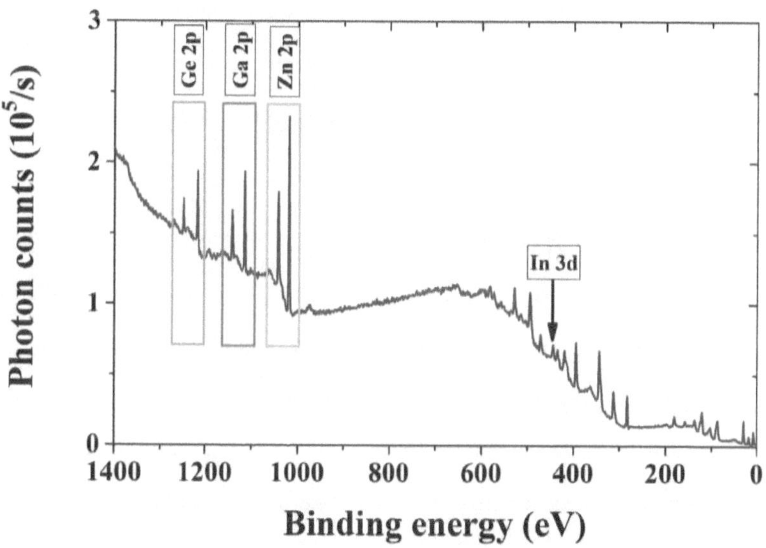

Figure 7.2 XPS survey spectra of a ZnGeN$_2$ film showing the presence of a substantial amount of Ga and In.

7.3.1.2 Unintentional Ga and In incorporation in ZnGeN₂ film

Figure 7.2 shows the XPS survey spectra of a $ZnGeN_2$ film grown on an InGaN/GaN/c-sapphire template (InGaN thickness ~ 60 nm). DEZn, GeH_4 and NH_3 were used as the precursors. N_2 was used as the carrier gas. No other gas was flowed during the growth. Based on the previously determined growth rates, the estimated thickness of the $ZnGeN_2$ film is 1 μm. The susceptor used in this growth had never been used for growth of any Ga, In or Sn containing materials. In the spectra shown in Fig. 7.2, Ge $2p$, Ga $2p$, Zn $2p$ and In $3d$ peaks are marked. No Sn peak was observed. The atomic composition of Zn, Ge, Ga and In determined were determined to be 53%, 16%, 23% and 8%, respectively. This data indicates that Ga and In preferentially replace Ge in the cation sublattice.

7.3.1.3 Reducing the cross-contamination

The dependence of cross-contamination on the history of the MOCVD reactor use was investigated. Figure 7.3 shows the XPS survey spectra of an InGaN film grown on a GaN/c-sapphire template. The MOCVD growth chamber was cleaned a few days before the growth of the sample in Fig. 7.3. No Zn, Ge, or Sn containing materials were grown (i.e., no Zn, Ge, or Sn precursor was flown) in the chamber in the meantime. The chamber cleaning is a routine maintenance procedure of the MOCVD system which includes cleaning of the inside chamber wall as well as the shower head.

In Fig. 7.3, the Ga $2p$ and In $3d$ peaks are marked. Unlike the XPS spectra in Fig. 7.1, which was from a InGaN films grown prior to cleaning the chamber, no Zn energy-level related peak could have been resolved which is evident from the zoomed in view of the spectra near the Zn $2p$ core-levels. This result indicates that a chamber cleaning procedure can effectively reduce cross-contamination at least to a level which can be probed by XPS.

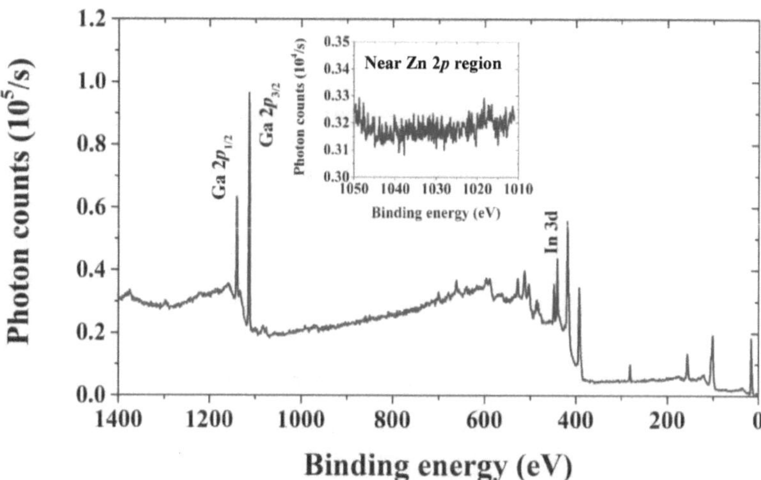

Figure 7.3 XPS survey spectra of an MOCVD grown InGaN films on a GaN. The sample was grown after a cleaning of the growth chamber. The inset shows the XPS spectra near the Zn $2p$ levels. No Zn core-level peak could have been resolved.

7.3.2 Implications of cross-contaminations

7.3.2.1 Contamination of the GaN or InGaN layers

Zn impurities in GaN are deep acceptors and form very efficient radiative centers giving rise to dominant blue-band luminescence, peaking between ~2.81 – 2.91 eV [109]. In addition, Zn-impurity related broad green, yellow, and red bands have also been reported [109]. In heavily Zn-doped GaN, these bands may dominate the near band edge emission

110

from GaN. Zn related emission in InGaN has also been investigated extensively [123, 124]. In fact, the first commercial blue LED was achieved by doping InGaN with Zn [125]. The energy levels of Zn in InGaN depends on the In content and the related emission peak shows a redshift with increase in In content [123]. For a high level of Zn impurity in InGaN QWs, the luminescence spectra of InGaN QWs can be dominated by the Zn-related bands [126] rather than by the band-to-band emission, which can be a potential way of achieving long visible wavelength emission from InGaN QWs. However, the FWHM of the Zn-related emission peak is usually much greater than the band-to-band emission peak [123, 126] which may not be useful for some applications, for example, full-color LED displays [127].

On the other hand, the group-IV elements Ge and Sn are possible shallow donors in GaN. The ionization energy of Ge and Sn, determined from the energy differences in PL bands, were reported to be 30 and 33 meV, respectively [128]. Unintentional Sn or Ge doping may result in doping of the active layer in GaN/InGaN/GaN QW LEDs. In the case of very high-levels of incorporation, the crystalline quality of the QW structure may degrade.

7.3.2.2 Contamination in $ZnGeN_2$ or $ZnSnN_2$

The atomic compositions determined from the XPS survey spectra in Fig. 7.2 indicate alloying of InGaN with $ZnGeN_2$. Alloys of $ZnGeN_2$ and GaN have been investigated both theoretically and experimentally [61, 110]. Based on first principles calculations, the bandgap of a 50/50 alloy of $ZnGeN_2$ and GaN is slightly larger than but within a few tens of meV of those of $ZnGeN_2$ and GaN [110]. One consequence of such alloying between ZnGeN$_2$ and GaN is the probable reduction in the band offsets with GaN as compared to pure ZnGeN$_2$ which has been experimentally demonstrated in Chapter 6 of this book.

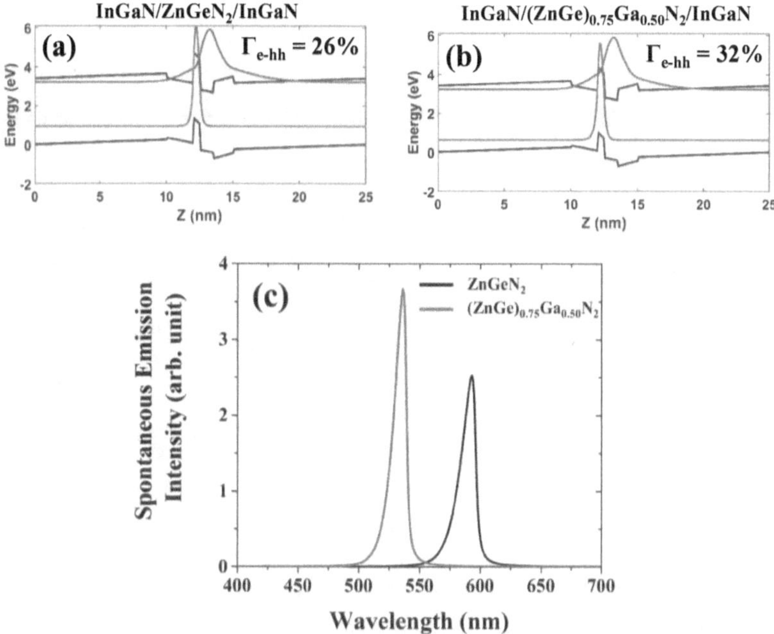

Figure 7.4 Band alignments in (a) a GaN/2 nm In$_{0.1}$Ga$_{0.9}$N/0.5 nm ZnGeN$_2$/1 nm In$_{0.1}$Ga$_{0.9}$N/1.5 nm Al$_{0.25}$Ga$_{0.75}$N/GaN QW and (b) a GaN/2 nm In$_{0.1}$Ga$_{0.9}$N/0.5 nm (ZnGe)$_{0.75}$Ga$_{0.50}$N$_2$-GaN/1 nm In$_{0.1}$Ga$_{0.9}$N/1.5 nm Al$_{0.25}$Ga$_{0.75}$N/GaN QW. (c) The spontaneous emission spectra of the QWs in (a) and (b).

The enabling factor for the improvement in the spontaneous emission properties of InGaN/ZnGeN$_2$ [32] and InGaN/ZnSnN$_2$ [35] heterostructure-based QWs as compared to that of conventional GaN/InGaN/GaN QWs is the large positive valence band offsets of $ZnGeN_2$ and $ZnSnN_2$, respectively, with GaN. We have investigated the effects of probable reduction in the valence band offset at the $ZnGeN_2$/GaN heterointerfaces due to the alloying of the Zn-IV-N_2 and III-Ns. Numerical simulation was performed to calculate the energy band alignment and the spontaneous emission properties of two InGaN/$(ZnGe)_{1-x}Ga_{2x}N_2$/InGaN heterostructre QWs with $x = 0$, and 0.25, respectively, using the similar technique described in Chapter 3.

Figure 7.4 (a, b) show the energy band alignments in a GaN/2 nm $In_{0.1}Ga_{0.9}N$/0.5 nm $ZnGeN_2$/1 nm $In_{0.1}Ga_{0.9}N$/1.5 nm $Al_{0.25}Ga_{0.75}N$/GaN QW and a GaN/2 nm $In_{0.1}Ga_{0.9}N$/0.5 nm $(ZnGe)_{0.75}Ga_{0.50}N_2$/1 nm $In_{0.1}Ga_{0.9}N$/1.5 nm $Al_{0.25}Ga_{0.75}N$/GaN QW, respectively. The electron-hole wavefunction overlap ($\Gamma_{e\text{-}hh}$) in the structure with $ZnGeN_2$-GaN alloy layer (Fig. 7.4 (b)) increases to 32% from 26% in the structure containing pure $ZnGeN_2$ layer (Fig. 7.4(a)). The conduction band minimum of $ZnGeN_2$ lies >1 eV above that of GaN, which distorts the electron wavefunction in InGaN/$ZnGeN_2$/InGaN structure. The reduction in the conduction band offset in $ZnGeN_2$-GaN alloy with GaN as compared to that of pure $ZnGeN_2$ makes the distortion of electron wavefunction less severe. Consequently, the electron-hole wavefunction overlap in the InGaN/$(ZnGe)_{0.75}Ga_{0.50}N_2$/InGaN enhances. The highest confined energy level in the valence band moves to a lower value in the structure with $(ZnGe)_{0.75}Ga_{0.5}N_2$ as compared to that in the structure with pure $ZnGeN_2$ layer. This results in a blue shift of the peak emission wavelength, which is shown in Fig. 7.4(c).

7.3.3 Compatibility of MOCVD growth conditions of InGaN and ZnGeN$_2$

Another major challenge is the difference in the growth temperature and reactor pressure required for desired In incorporation in InGaN with those required for growth of high crystalline quality ZnGeN$_2$. In Chapter 5, it was shown that the ZnGeN$_2$ films grown at 600 °C and 650 °C suffers from high densities of extended defects as compared to those grown at 735 °C and 770°C. Therefore, it is desirable to grow the InGaN QWs at higher temperatures. However, we found that, in our MOCVD chamber, In incorporation in InGaN is very low at this range of growth temperature causing a trade-off between quality of ZnGeN$_2$ and In composition. On the other hand, higher reactor pressure is required to achieve stoichiometric ZnGeN$_2$ at higher growth temperature whereas lower reactor pressure helps increasing the In incorporation in InGaN.

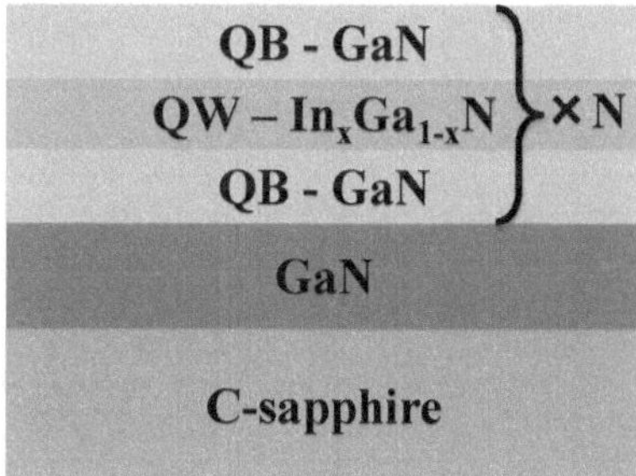

Figure 7.5 Schematic of a conventional GaN/InGaN/GaN QW structure.

7.4 Calibration of MOCVD growth of conventional GaN/InGaN/GaN QWs

The effects of growth parameters including the substrate temperature, reactor pressure, NH$_3$ molar flow rate and growth rate on In incorporation are well documented in the relevant literature. However, it is well known that every growth reactor is different and require fine tuning of the growth parameters for achieving optimum results. Therefore, we have performed a systematic calibration of growth conditions for conventional GaN/InGaN/GaN QWs. Figure 7.5 shows the schematic of a conventional InGaN QW structure with GaN barriers. Three samples having the structure in Fig. 7.5 (N = 3) were grown using different combinations of growth temperature (T), reactor pressure (P) and TMIn flow rate. Table 7.1 lists the growth conditions for these three samples.

Table 7.1 Growth temperature (T) in °C, reactor pressure (P) in Torr, triethylgallium (TEGa) and trimethylindium (TMIn) flows in μmol/min, ammonia (NH_3) flow in sLm, and duration (t) in sec used for growing the GaN barrier and InGaN well layers for three conventional GaN/InGaN/GaN QW samples (A, B, and C). The growth IDs of the samples are mentioned below their names in the leftmost column.

Sample (Growth ID)	# of QW	Barrier						QW					
		T (°C)	P (Torr)	TEGa (μmol/min)	NH_3 (sLm)	t (sec)	T (°C)	P (Torr)	TEGa (μmol/min)	TMIn (μmol/min)	NH_3 (sLm)	t (sec)
A (QW072)	3	640	500	9	3	230	640	500	9	7	3	80
B (QW076)	3	600	500	9	3	230	600	500	9	8	3	80
C (QW078)	3	600	200	9	3	230	600	200	9	8	3	80

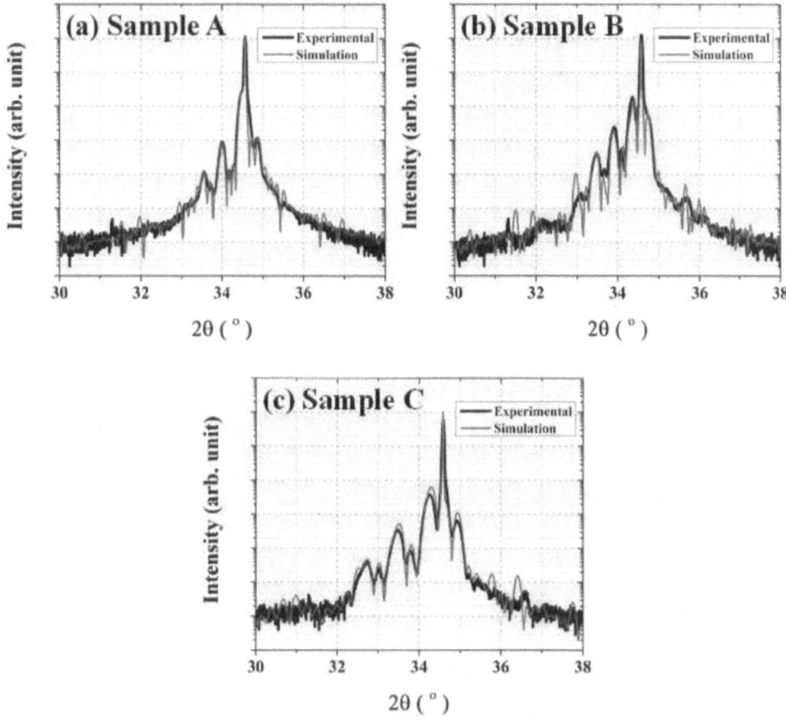

Figure 7.6 (a-c) XRD 2θ-ω scan profiles (thick black curve) obtained from samples A, B, and C, respectively. The fitted curves (thin red curves) generated using Bruker LEPTOS [129] software is shown in each case.

XRD 2θ-ω scan profiles were obtained from samples A, B, and C to investigate the structural properties of the grown samples and are shown in Fig. 7.6(a-c), respectively, by thick black curves. With a view to estimating the In content and thickness of the barrier and well thicknesses, simulated data were fitted over the experimental ones using Bruker

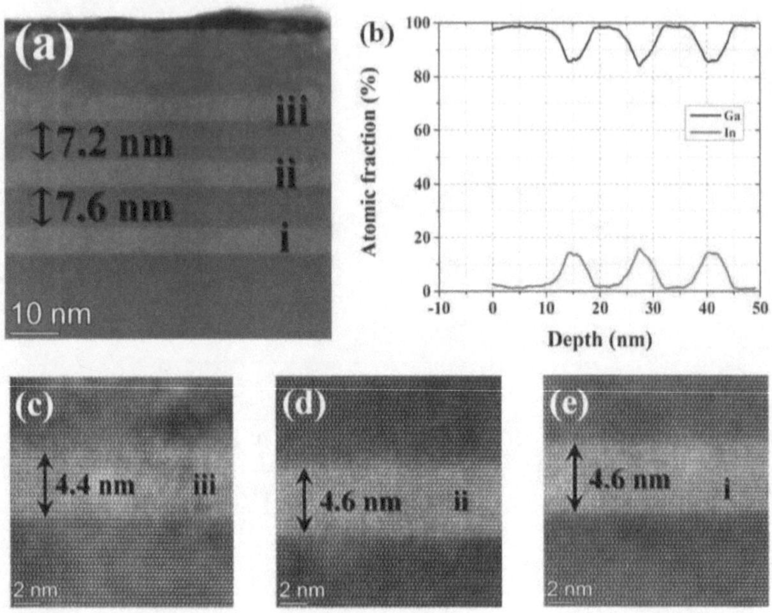

Figure 7.7 (a) Cross-sectional STEM image of sample C showing three QWs structure. (b) EDS line scan along the cross-section in (a) showing the atomic mole fractions of Ga and In in the barrier and the well region. (c-e) High magnification STEM images of the top, middle, and bottom wells, respectively, in (a).

LEPTOS software [129] for each sample. The fitted 2θ-ω profiles are shown by thin red curves. The barrier thickness, well thickness and In content values extracted from fitting were 13.2 nm, 5.7 nm, and 6.3% for sample A, 11.5 nm, 8.6 nm, and 9.4% for sample B, and 7 nm, 4.9 nm, and 14.1% for sample C. A gradual increase in In content from sample A to sample C is consistent with the variation in growth conditions for these samples.

To further investigate the structural properties of the grown samples, STEM imaging was performed on sample C. Figure 7.7 (a) is the cross-section STEM image of sample C showing the alternative GaN barrier and InGaN well layers. The bottom, middle and top QWs are marked as i, ii, and iii, respectively. Figure 7.7(c-e) shows high-magnification STEM images of the top, middle, and bottom QWs, respectively. While the bottom GaN/InGaN interface is atomically smooth in all three wells, the top InGaN/GaN interface become slightly rough in the top quantum (iii) well (Fig. 7.7 (c)). The average barrier and well thicknesses were estimated to be ~7.4 nm and 4.6 nm. Figure 7.7 (b) plots EDS line scans obtained along the cross-section in Fig. 7.7(a). The atomic mole fraction in the well regions is ~14%. The barrier and well thickness determined from STEM images as well as the In composition determined from the EDS lines scan agrees well with corresponding values determined form the XRD 2θ-ω scan profile.

Figure 7.8 (a) presents the room temperature CL spectra obtained from sample A, B, and C. A beam acceleration voltage (V_{beam}) of 5 kV and a beam current (I_{beam}) of 0.71 nA were used for all three measurements. The spectra were normalized with respect to the corresponding maximum intensity. Each spectrum in Fig. 7.8 (a) shows two peaks – peak 1 (marked by orange stars) and peak 2 (marked by green disks). The GaN near band edge (NBE) peaks are also present in all the spectra. For samples A, B, and C, the positions of peaks 1 are 391 nm (3.18 eV), 414 nm (3.00 eV), and 470 nm (2.64 nm), respectively, whereas the positions of peaks 2 are 458 nm (2.71 eV), 491 nm (2.52 eV), and 2.34 nm (2.34 eV), respectively. These values are listed in Table 7.2.

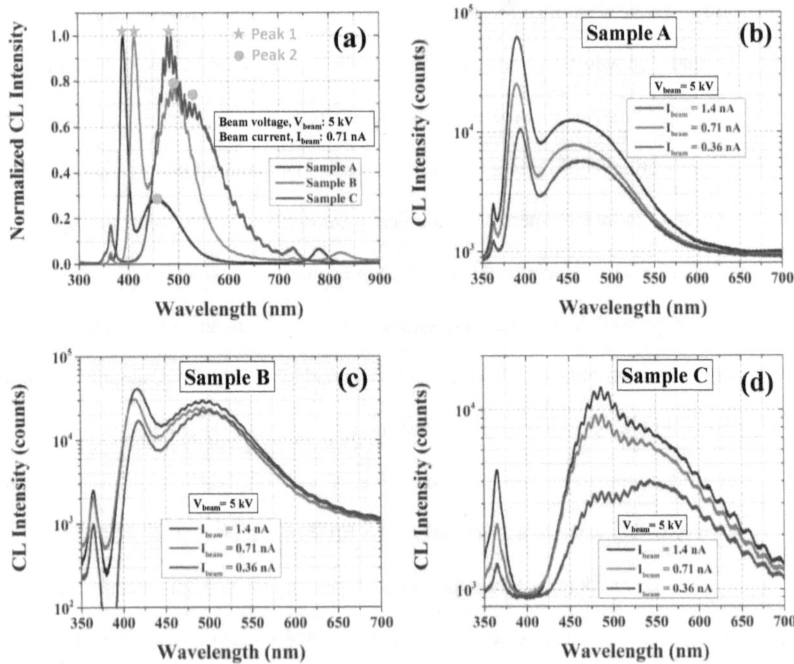

Figure 7.8 (a) Room temperature CL spectra of samples A, B, and C. A 5 kV electron beam volatge and 0.71 nA beam current was used. (b-d) the CL spectra obtained with three different electron beam currents (0.36 nA, 0.71 nA, and 1.4 nA) from sample A, B and C, respectively.

Table 7.2 Positions of peak 1 and peak 2 in samples A, B, and C along with the separation between them.

Sample	Position of peak 1		Position of peak 2		Separation between peaks 1 and 2 (eV)
	Wavelength (nm)	Energy (eV)	Wavelength (nm)	Energy (eV)	
A	391	3.18	458	2.71	0.47
B	414	3.00	491	2.52	0.48
C	470	2.64	531	2.34	0.3

The separations between peak 1 and peak 2 are 0.47 eV, 0.48 eV, and 0.30 eV for samples A, B, and C, respectively. Based on the well thickness and In content determined from XRD, STEM, and EDS, peaks 1 correspond to the band-to-band emissions from the InGaN well in sample A, B, and C. On the other hand, based on the separation from peak 1 and the peak FWHM values, peak 2 in each sample can be assigned to the Zn-related emission [123]. Unintentional Zn incorporation in these samples is a result of cross-contamination which has already been discussed in section 7.2.

Figures 7.8(b-d) show room temperature CL spectra obtained from samples A, B, and C, respectively, with three different beam current values (0.36 nA, 0.71 nA, and 1.4 nA). The beam acceleration voltage was set at 5 kV for each measurement. For each of the samples, the relative intensity of peak 1 to peak 2 increases with increase in beam current which indicates the saturation of recombination centers that gives rise to peak 2. This further supports the assignment of peak 1 to excitonic emission and peak 2 to Zn-related emission [130].

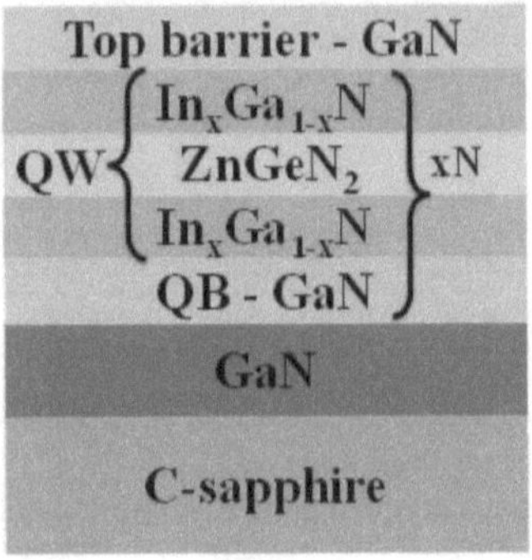

Figure 7.9 Schematic of the GaN/InGaN/ZnGeN$_2$/InGaN/GaN QW structures.

Table 7.3 Growth temperature (T) in °C, reactor pressure (P) in Torr, metalorganic precursor flow rates (TEGa and/or TMIn) in μmol/min, NH$_3$ flow in sLm and growth duration (t) in sec for the barrier and well layers used for two conventional GaN/InGaN/GaN QW (sample D, and H) structures. The growth IDs of the samples are also mentioned in the leftmost column below the respective sample name.

Sample (Growth ID)	# of QW, N	Barrier, GaN					Well, InGaN					
		T (°C)	P (Torr)	TEGa (μmol/min)	NH$_3$ (sLm)	t (sec)	T (°C)	P (Torr)	TEGa (μmol/min)	TMIn (μmol/min)	NH$_3$ (sLm)	t (sec)
D (QW0027)	4	780	400	10	4	420	680	400	7	23	4	180
H (QW0092)	3	735	500	8	3	230	735	500	9	8	3	50

7.5 MOCVD growth of GaN/InGaN/ZnGeN$_2$/InGaN/GaN QW structures

A range of GaN/InGaN/ZnGeN$_2$/InGaN/GaN QW structures was grown. The general structure of the samples is shown by the schematic in Fig. 7.9. The luminescence spectra of the GaN/InGaN/ZnGeN$_2$/InGaN/GaN QW samples are compared with those of conventional GaN/InGaN/GaN QWs. The structure of the conventional QWs is similar to that shown in Fig. 7.9 minus the ZnGeN$_2$ layer which is shown in Fig. 7.5.

Table 7.3 lists the growth conditions for the barrier and well layers of two GaN/InGaN/GaN QW structures (sample D, and H) whereas Table 7.4 lists the same for seven GaN/InGaN/ZnGeN$_2$/InGaN/GaN QW structures (samples E-G and I-L). The growth conditions for the ZnGeN$_2$ layers in samples E-G as well as samples I-L are listed in Table 7.5. For samples E-G and I-L, the same growth conditions were used for two InGaN layers in a period (below and above the ZnGeN$_2$ layer). The growth temperature of the ZnGeN$_2$ was 680 °C for samples E-G and 735 °C for sample I-L.

For sample E-G, the growth durations of two InGaN layers in a period was kept the same. For sample E, the growth conditions for the GaN and InGaN layers were the same as those of the conventional GaN/InGaN/GaN QW sample D. Besides, the sum of the growth duration of the InGaN layers below and above the ZnGeN$_2$ layers was kept same. In sample E, the InGaN layers were grown using the same conditions as sample E but the growth durations of the InGaN layers were reduced by 33.3%. Sample G is same as the sample E except 26% less TMIn flow rates were used for growing the InGaN layers. The GaN and InGaN layers of sample E were grown under identical conditions as the reference sample D. Therefore, one would expect that a conventional GaN/InGaN/GaN QW grown using the same conditions and durations for the GaN and InGaN layers used for sample F will have a similar In content but results in a narrower well than sample D. Similarly, a conventional GaN/InGaN/GaN QW grown using the same conditions and durations for the GaN and InGaN layers used for sample G will have a similar well thickness but smaller In content than sample D. The growth conditions for the $ZnGeN_2$ layer were chosen to obtain stoichiometric $ZnGeN_2$ and were kept the same for sample E-G.

Table 7.4 Growth temperature (T) in °C, reactor pressure (P) in Torr, metalorganic precursor flow rates (TEGa and/or TMIn) in μmol/min, NH_3 flow in sLm and growth duration (t) in sec for the barrier and well layers used for two sets of GaN/InGaN/$ZnGeN_2$/InGaN/GaN QW structures. The reference conventional GaN/InGaN/GaN QW structures are sample D for samples E-G and sample H for samples I-L. The growth IDs of the samples are also mentioned in the leftmost column below the respective sample name.

| Sample (Growth ID) | Ref. | # of QW, N | Barrier, GaN | | | | | Well, InGaN | | | | | | |
			T (°C)	P (Torr)	TEGa (μmol/min)	NH_3 (sLm)	t (sec)	T (°C)	P (Torr)	TEGa (μmol/min)	TMIn (μmol/min)	NH_3 (sLm)	t (sec) Below ZGN	t (sec) Above ZGN
E (QW0029)		4	780	400	10	4	420	680	400	7	23	4	90	90
F (QW0030)	D	4	780	400	10	4	420	680	400	7	23	4	60	60
G (QW0031)		4	780	400	10	4	420	680	400	7	17	4	90	90
I (QW0098)		3	735	500	9	3	230	735	500	9	8	3	30	20
J (QW0095)	H	3	735	500	9	3	230	735	500	9	8	3	30	20
K (QW0097)		3	735	500	9	3	230	735	500	9	8	3	30	20
L (QW0100)		3	735	500	9	3	230	735	500	9	8	3	18	32

Table 7.5 Growth temperature (T) in °C, reactor pressure (P) in Torr, diethylzinc (DEZn) flow rate in (μmol/min), germane (GeH$_4$) flow rate in μmol/min, ammonia (NH$_3$) flow rate in sLm and growth duration (t) in sec used for the ZnGeN$_2$ layers in samples E-G and samples I-L.

Sample	ZnGeN$_2$					
	T (°C)	P (Torr)	DEZn (μmol/min)	GeH$_4$ (μmol/min)	NH$_3$ (sLm)	t (sec)
E, F, G	680	400	340	4.5	4	5
I	735	500	340	4.5	3	3.5
J	735	500	340	4.5	3	5
K	735	500	340	4.5	3	7
L	735	500	340	4.5	3	3.5

STEM imaging was carried out to characterize the resulting structures. Figure 7.10 is the cross-sectional STEM image of sample F, showing four QWs. The total thickness of the InGaN/ZnGeN$_2$/InGaN heterostructures was close to 10.5 nm, whereas the barriers were ~50 nm thick. The STEM image clearly shows a thin layer of different contrast inside each of the InGaN well layers. Based on the growth recipe, these thin layers are expected to be ZnGeN$_2$.

Figure 7.11 shows the room temperature CL spectra measured from the GaN/InGaN/ZnGeN$_2$/InGaN/GaN QW samples E-G along with that from the reference conventional GaN/InGaN/GaN QW sample D. The inset shows the same spectra but normalized with respect to the corresponding maximum intensities. The CL spectrum of the reference sample D shows two obvious peaks – one at 400 nm and the other around

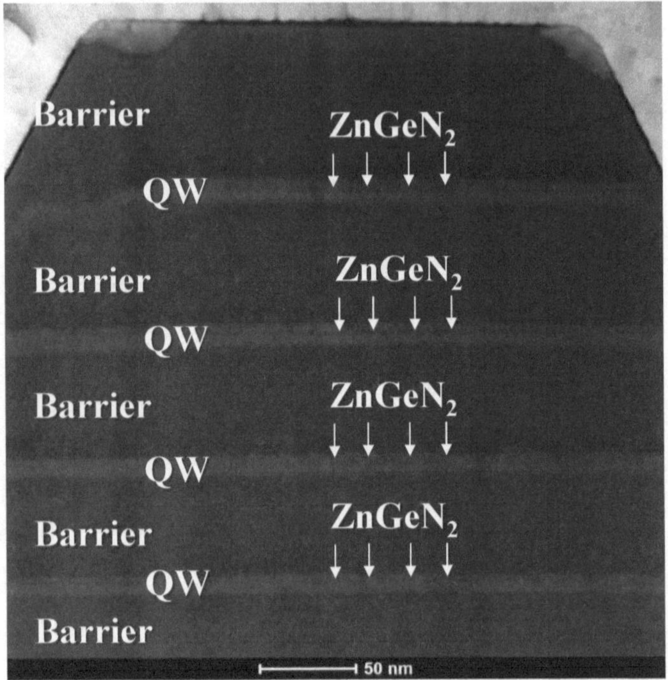

Figure 7.10 Cross-sectional STEM image of GaN/InGaN/ZnGeN₂/InGaN/GaN QW sample F. The thin ZnGeN₂ layers are marked by white arrows.

~470 nm which, based on the discussion in section 7.3, correspond to the band-to-band emission and Zn-related emission from the InGaN well layers. The maximum intensity for samples E, F, and G is drastically reduced as compared to sample D. For example, sample E, which was grown with identical conditions for barrier and well layers as the reference sample D, the room temperature CL intensity is more than one order of magnitude lower than that of sample D. One probable cause of the reduction in CL intensity in samples E-F

125

as compared to the reference sample D is the structural degradation in the QW, for example, increase in interfacial roughness and extended defect density which act as non-radiative recombination centers, as a result of inserting the $ZnGeN_2$ layers. As shown in chapter 4, the density of extended defects in $ZnGeN_2$ decreases with increase in growth temperature. Therefore, an increase in growth temperature from 680 °C, which was used for samples E-G, may help to improve the emission properties in terms of the intensity of the $GaN/InGaN/ZnGeN_2/InGaN/GaN$ QWs.

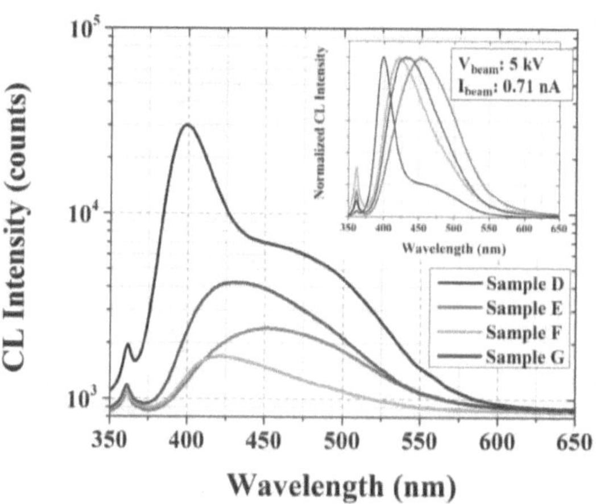

Figure 7.11 Room temperature CL spectra obtained from samples D-G. Beam voltage and current values were set to 5 kV and 0.71 nA, respectively. The inset shows the spectra normalized with respect to corresponding maximum intensity.

Besides, the inset in Fig. 7.11 clearly shows a red shift in the peak position as well as peak-broadening in sample E as compared to the band-to-band emission peak of the reference sample D. A peak red shift in the GaN/InGaN/ZnGeN$_2$/InGaN/GaN QW structure as compared to the conventional GaN/InGaN/GaN QW structure with identical InGaN thickness and In compositions has been predicted by numerical simulations [32]. However, the peak in sample E is very close to that of the Zn-related emission in peak in sample D. Therefore, it is not reasonable to attribute the observed peak shift in sample E from sample D to the narrowing of the effective band gap by insertion of ZnGeN$_2$ layer. Note that the peak position in samples F and G are not directly comparable to that in sample D since sample F has a narrower InGaN well thickness and sample G has smaller In content as compared to the reference sample D.

The GaN/InGaN/ZnGeN$_2$/InGaN/GaN QW samples I-L were grown using growth conditions of the conventional GaN/InGaN/GaN QW sample H as reference. A 735 °C growth temperature, and a 500 Torr reactor pressure and a 9 µmol/min of TEGa flow rate were used for both the GaN barrier and InGaN well layers in all five samples (H-L). The TMIn flow rate used for the InGaN well layers was 8 µmol/min for all five of these samples (H-L). In addition, the sum of the growth durations of two InGaN layers in a period (below and above the ZnGeN$_2$ layer) of samples I-L was equal to the growth duration of the InGaN layer in sample H. The growth durations for the InGaN layers below and above the ZnGeN$_2$ layers in a period were 30 sec and 20 sec, respectively, for samples I-K whereas 18 sec and 32 sec, respectively, for sample L. The growth conditions used for the ZnGeN$_2$ layers in samples I-L are listed in Table 7.5. The growth temperature, reactor pressure, DEZn flow rate, GeH$_4$ rate and NH$_3$ flow rate were 735 °C, 500 Torr, 340 µmol/min, 4.5 µmol/min and 3 sLM, respectively. The growth durations of the ZnGeN$_2$ layer were 3.5 sec for samples I and L, 5 sec for sample J, and 7 sec for sample K.

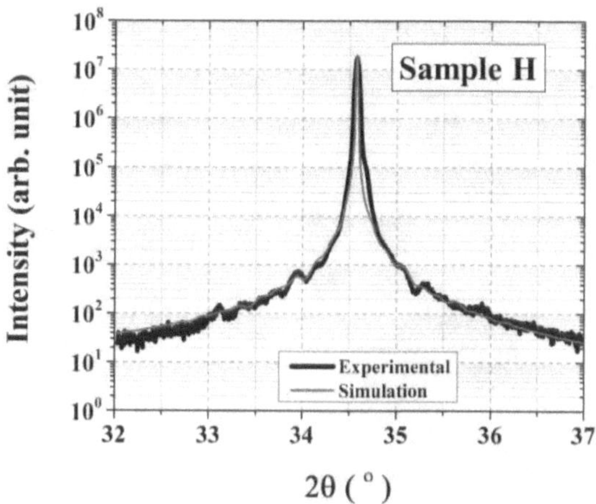

Figure 7.12 XRD 2θ-ω scan profile of the conventional GaN/InGaN/GaN QW sample H. The black curve is the experimental data whereas the red curve was obtained by fitting of the simulated profile using the Bruker LEPTOS software. The estimated barrier thickness, well thickness and In content from the fitting were 13 nm, 5 nm and 0.35%.

Figure 7.12 shows the XRD 2θ-ω scan profile (thick black curve) obtained from the conventional GaN/InGaN/GaN QW sample H. The simulated XRD data from a model sample was fitted over the experimental data using the Bruker LEPTOS software [129]. The estimated barrier thickness, well thickness and In content values were 13 nm, 5 nm and 0.35%.

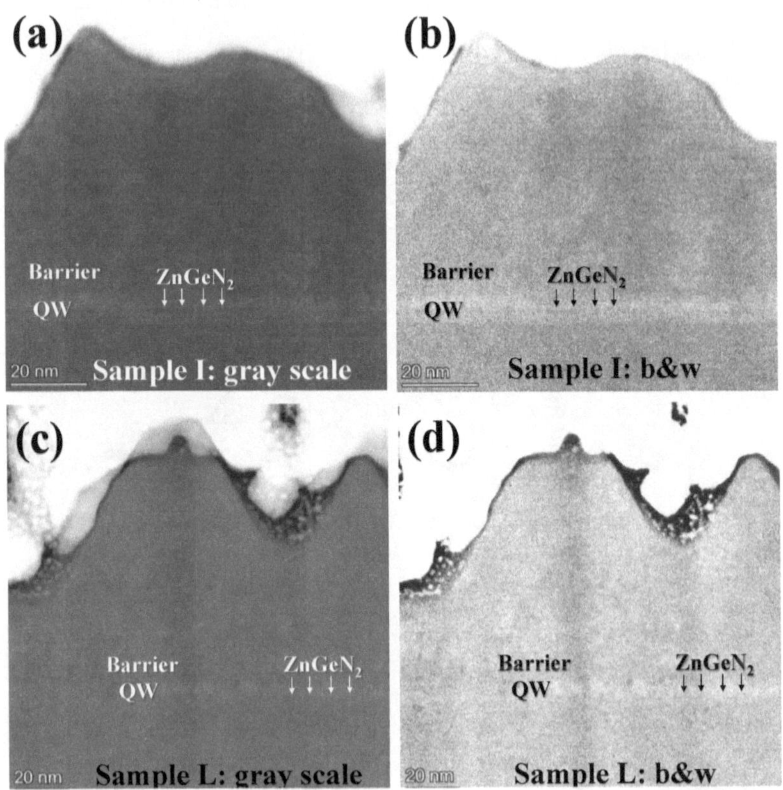

Figure 7.13 Cross-sectional STEM image of the GaN/InGaN/ZngeN$_2$/InGaN/GaN QW sample I (a, b) and sample L (c, d) shown in gray scale (a, c) and in black-and-white for better visibility of the well region (b, d). The ZnGeN$_2$ layers are marked by downward arrows. The low contrast difference between GaN and InGaN layer is due to the very low In content in the well. Although the sample is supposed to have three QWs, only one can be noticed.

Figures 7.13(a) and (c) show the cross-sectional STEM image (in gray scale) of a GaN/InGaN/ZnGeN$_2$/InGaN/GaN QW structures sample I and sample L, respectively. For better visibility of the well layer, the same image is shown in black-and-white in Figs. 7.13 (b) and (d), respectively. The low contrast difference between the barrier and the well layer is due to the low In content in the well. For both samples, among the three QW periods in the sample, only one can be visually resolved.

Figure 7.14(a) presents the room temperature PL spectra of the conventional GaN/InGaN/GaN QW sample H along with GaN/InGaN/ZnGeN$_2$/InGaN/GaN QW samples I-L. All the spectra show a peak at 363-364 nm (3.41 eV), which is near the band edge of GaN. The PL spectrum of Sample H (black curve) shows a dominant peak 432 nm (2.88 eV). The well-known 'yellow-band' related emission peak is also obvious for this sample. For sample I, the PL spectrum (red curve) shows two overlapping emission band with peaks at 455 nm (2.73 eV), and 479 nm (2.59 eV). The 'yellow-band' peak is not obvious for this sample. The PL spectra of sample J (green curve) shows a clear peak at 445 nm (2.79 eV) and a shoulder at ~472 nm (2.63 eV). For sample K, only the 'yellow-band' peak was observed (blue curve). The room temperature PL spectrum of sample L (magenta curve) shows a dominant peak at 512 nm (2.43 eV) and one shoulder at around 461 nm (2.7 eV). The peaks at 432 nm (2.88 eV), 455 nm (2.73 eV), 445 nm (2.79 eV), and 461 nm (2.7 eV) form samples H, I, J, and L, respectively, are marked as P1 in the figure whereas those at 479 nm (2.59 eV), 472 nm (2.63 eV), and 512 nm (2.43 eV), from samples I, J, and L, respectively, are marked as P2 in the figure. Table 7.6 lists the peak positions in the PL spectra form samples H-L.

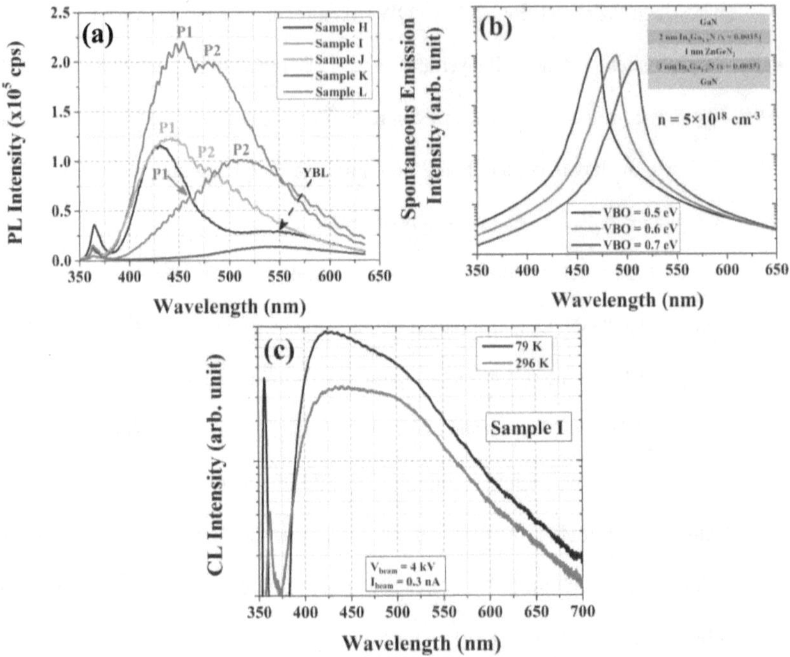

Figure 7.14 (a) Room temperature PL spectra of samples H-L. (b) Calculated spontaneous emission spectra of a GaN/3 nm $In_{0.0035}Ga0._{9965}N$/1 nm $ZnGeN_2$/ 2 nm $In_{0.0035}Ga0._{9965}N$/ GaN QW structure. To account for the likely contamination of $ZnGeN_2$ by Ga, which results in an alloying of $ZnGeN_2$ and GaN, the valence band offset (VBO) of the $ZnGeN_2$ layer with GaN was set to at lower values than that of pure $ZnGeN_2$. Spectra calculated for approximated three VBO values are shown. (c) CL spectra of sample I measured at 296 K as well as at 79 K.

Table 7.6 Positions of the GaN near band edge peak, Zn-related emission peak P1, unassigned but probably InGaN/ZnGeN$_2$/InGaN band-to-band emission peak P2 and the 'yellow band' emission peak in the room temperature PL spectra of sample H-L.

Sample	Peak positions							
	GaN NBE peak		Peak P1		Peak P2		Yellow band	
	Wavelength (nm)	Energy (eV)	Wavelength (nm)	Energy (eV)	Wavelength (nm)	Energy (eV)	Wavelength (nm)	Energy (eV)
H	365	3.41	432	2.88	Not observed		550	2.26
I	363	3.42	455	2.73	479	2.59	Not obvious	
J	363	3.42	445	2.79	472	2.63	Not obvious	
K	363	3.42	Not observed		Not observed		550	2.26
L	364	3.41	461	2.70	512	2.43	Not obvious	

For sample H, the band-to-band emission peaks from the QW cannot be resolved from near band-edge peak of GaN (~3.41 eV) due to the very low In content (0.35% determined form XRD) in the InGaN well layer. The position of the peak P1 (2.88 eV) in PL spectra of sample H lies within the range of reported Zn-related emission peak in GaN [109]. The unintentional Zn incorporation in III-N films grown in our MOCVD reactor has already been discussed in section 7.2.1. The peaks P1 in the PL spectra of samples I, J, and L are also attributed to Zn-related emission. One possible reason for the variation in the position of the Zn-related emission peak (P1) from sample I, K, and L is slight variation in In content in this samples.

The possibility of the peak P2 to be related to the band-to-band emission in the InGaN/ZnGeN$_2$/InGaN layer was considered. Based on numerical simulations [32], the spontaneous emission peak of a GaN/InGaN/ZnGeN$_2$/InGaN/GaN QWs structure is red shifted as compared to that of a conventional GaN/InGaN/GaN QW structure with same In content and total InGaN layer thickness. With a view to evaluating the possibility of peak P2 in PL spectra of samples I, J, and L to be caused by hole confinement in the ZnGeN$_2$ layer, numerical simulation of a GaN/3 nm In$_x$Ga$_{1-x}$N/1 nm ZnGeN$_2$/2 nm In$_x$Ga$_{1-x}$N/GaN structure was performed using the same technique described in Chapter 3. The In composition (0.35%) and the sum of the two InGaN layer thicknesses (5 nm) were same as the In content and InGaN layer thickness of sample H, respectively, estimated from XRD. Carrier concentrations were assumed to be 5×10^{18} cm^{-3}. The thicknesses of the individual InGaN layers were selected based on the growth duration of these layers in sample I. To account for the likely contamination of the ZnGeN$_2$ with Ga, which was shown to result in a ZnGeN$_2$-GaN alloy, the valence band offsets of the ZnGeN$_2$ layer with GaN were approximated to be lower than that of pure ZnGeN$_2$. Three simulations were performed for three different values of the valence band offset between the GaN and the ZnGeN layers. Figure 7.13(b) shows the calculated spontaneous emission spectra for valence band offset values of 0.5 eV, 0.6 eV, and 0.7 eV for which the peak intensity occurred at 472 nm, 488 nm, and 508 nm, respectively. These values, especially, the 472 nm and 488 nm are reasonably close to the position of peak P2 from sample I (479 nm). By comparing the simulated spectra with the measured PL spectra of sample I, one may qualitatively argue that the peak P2 in the measured PL spectra is related to the band-to-band emission in the InGaN/ZnGeN$_2$/InGaN active layers which is red shifted by ~0.82 eV as compared to the band-to-band emission in the InGaN active layer in the reference conventional QW (sample H). However, additional experimental evidence is necessary to support this claim, which is an ongoing work.

Low temperature CL measurements were performed on sample I. Figure 7.13(c) shows the CL spectra obtained at 79 K as well as 296 K. The 296 K CL spectrum appears to be an envelope of two peaks at ~441 nm (2.81 e) and ~500 nm (2.49 eV). These two probably correspond to peaks P1 and P2, respectively, in the room temperature PL spectra of sample I (red curve in Fig. 7.13(a)). At 79 K, the intensities of both peaks increased by a factor of 2-3, but the peak P1 became dominant. The integrated CL intensity over 380 – 600 nm at 296 K is 47% of that at 79 K.

The maximum PL intensity decreases from sample I to sample J to sample K. The only difference among these samples is the ZnGeN$_2$ layer growth duration. Based on the numerical simulation, an increase in the ZnGeN$_2$ layer thickness in the GaN/InGaN/ZnGeN$_2$/InGaN/GaN QW results in (i) a redshift of the spontaneous emission peak, and (ii) a reduction in the predicted electron-hole wavefunction overlap ($\Gamma_{e\text{-}hh}$) [131]. The correlation of the emission intensity and the ZnGeN$_2$ layer growth duration in samples I-K is qualitatively consistent with the numerical simulation results. However, no significant shift of peak P2, which is expected to be related to the band-to-band emission related peak from InGaN/ZnGeN$_2$/InGaN active layer, between samples I and J is apparent in Fig. 7.13 (a). Additional investigation is necessary for the understanding of the physical origin the of reduction in the overall emission intensity with increase in the ZnGeN$_2$ layers growth duration.

The peak P2 in the PL spectrum of sample L is ~ 0.16 eV red shifted with respect to the same peak from sample I. The only difference between samples I and L is the growth durations of the InGaN layers below and above the ZnGeN$_2$ which is expected to result in different positions of the ZnGeN$_2$ layer in the inside InGaN well. Based on numerical simulations [131], such a shift in the position of the ZnGeN$_2$ layer inside the InGaN layer would red-shift the spontaneous emission peak. Therefore, the experimental observations are consistent with the predictions by numerical simulation.

7.6 Conclusions

GaN/InGaN/ZnGeN$_2$/InGaN/GaN heterostructure QW structures were grown by MOCVD. The growth rates of the GaN and InGaN layers as well as the In content in the InGaN were calibrated by growing conventional GaN/InGaN/GaN QWs. Two sets of samples were prepared with two different growth temperatures (680 °C and 735 °C) of ZnGeN$_2$ layers. The structures were confirmed from STEM images which clearly showed thin ZnGeN$_2$ layers sandwiched between two InGaN layers. The emission properties of the GaN/InGaN/ZnGeN$_2$/InGaN/GaN QW structures were investigated using CL and PL spectroscopy measurements, which showed dependence on the ZnGeN$_2$ layer growth temperature, the thickness of the ZnGeN$_2$ layer and the position of the ZnGeN$_2$ layer (i.e., the relative thickness of the InGaN layers below and above the ZnGeN$_2$ layer). For a 685 °C well growth temperature, the emission intensity of GaN/InGaN/ZnGeN$_2$/InGaN/GaN QWs was substantially reduced as compared to the conventional GaN/InGaN/GaN QWs grown with same conditions. For a 735 °C growth temperature, the PL and CL spectra of GaN/InGaN/ZnGeN$_2$/InGaN/GaN QWs showed the emergence of a new peak which was not present in the spectra of the reference conventional GaN/InGaN/GaN QWs. The position of this peak is consistent with the position of the spontaneous emission peak of a similar structure calculated using numerical simulation. However, additional experimental evidence is required to completely understand the emission properties of InGaN/ZnGeN$_2$/InGaN heterostructures quantum wells.

Chapter 8 Conclusions and future work

8.1 Conclusions

The major contributions of this book can be summarized as follows.

First, a design of $InGaN/ZnSnN_2$ heterostructure based QW structure has been investigated. The major advantages of inserting a thin $ZnSnN_2$ layer inside InGaN QWs were (i) achieving a targeted emission wavelength with much less In composition, and (ii) substantial enhancement of the electron-hole wave function overlap as compared to conventional InGaN QWs. Both improvements were attributed to the strong hole confinement in the $ZnSnN_2$ layer due to the large valence band offset. When designed for a 600 nm peak emission wavelength, a $InGaN/ZnSnN_2$ heterostructure QW requires only one third of the In content as compared to that of the conventional InGaN QW with similar thickness. The peak spontaneous emission intensity and the spontaneous emission radiative recombination rate of this $InGaN-ZnSnN_2$ QW were enhanced by two orders of magnitude enhancement from those of the conventional InGaN QW. The proposed $InGaN/ZnSnN_2$ QW structure is promising to address the current efficiency challenge in InGaN QW LEDs emitting beyond blue and green.

The remaining part of the book has focused on MOCVD growth and characterization of $ZnGeN_2$ thin films and GaN/InGaN/$ZnGeN_2$/InGaN/GaN heterostructure QWs.

$ZnGeN_2$ films were grown on sapphire substrates as well as on GaN/c-sapphire templates using MOCVD. Single crystalline $ZnGeN_2$ films with (001) out of plane orientation and planar surface morphology were grown on c- and a-sapphire substrates whereas the single crystalline $ZnGeN_2$ films on r-sapphire substrates grew along [010] direction and showed stepped surface morphology. Room temperature PL spectra of the films showed only deep-level defect related peaks. Room temperature PLE spectra showed peaks around 3.4 eV, which is close to the predicted band gap of $ZnGeN_2$. Films were found to show n-type conductivity with carrier concentrations in the range of 2×10^{18} – 2×10^{19} cm^{-3}. Hall electron mobilities up to 17 $cm^2/V\cdot s$ were measured at room temperature. The effect of thermal annealing as studied on $ZnGeN_2$ films grown on c-sapphire substrates. Room temperature Raman spectroscopy and PLE spectroscopy indicated thermal annealing induced an increase in cation ordering in $ZnGeN_2$ films.

In addition, $ZnGeN_2$ films were grown on GaN/c-sapphire templates and the effects of growth parameters on cation compositions were investigated. The Zn/(Zn+Ge) composition in the films decreased with increase in growth temperature but increased with increase in total reactor pressure and ratio of Zn precursor flow rate to that of Ge. Atom probe tomography measurements indicated homogenous distributions of the cations with out significant clustering. The crystalline quality and surface morphology of the films were found to depend on the cation compositions. Near stoichiometric films were found to be single crystalline whereas films with off-stoichiometric Zn/(Zn+Ge) compositions were not single crystalline. The former showed smooth planar surface morphology with decent roughness whereas the surface morphology of the latter showed either the presence of crystallites (zinc-rich) films) or faceting (zinc-poor). Room temperature Raman spectra indicated the presence of cation disorder. The measured CL and PL luminescence peaks show PL from the $ZnGeN_2$ layer reminiscent of the "yellow band" PL commonly observed in GaN.

Next, experimental determinations of valence band offset of $ZnGeN_2$ and $(ZnGe)_{0.94}Ga_{0.12}N_2$ with GaN were performed for the first time to the best of our knowledge. The measured valence band offset at the $ZnGeN_2$/GaN heterointerface (1.45 eV - 1.65 eV) is comparable to the predicted value from first-principles calculations using explicit interface calculations. For the $(ZnGe)_{0.94}Ga_{0.12}N_2$/GaN heterointerface, the VBO was measured to be 1.29 eV, which is very close to the predicted value from the theoretically calculated VBO of $ZnGeN_2$ assuming a linear dependence of VBO on composition.

Finally, work on implementation of $GaN/InGaN/ZnGeN_2/InGaN/GaN$ QW structures was presented. The technical challenges to implementation of such heterostructures including the cross-contamination and its implications on emission properties of the devices were discussed. The structures of the grown samples were verified using STEM imaging and XRD measurement. The PL and CL spectra of the $GaN/InGaN/ZnGeN_2/InGaN/GaN$ heterostructure QWs showed emergence of new peaks or shift in peak positions when compared with spectra of reference conventional quantum probably due to the modification of band alignment in the heterostructure QW structures. The emission intensities and peak positions in the PL and CL spectra of GaN/InGaN/ZnGeN$_2$/InGaN/GaN heterostructure QWs showed dependence on the thickness and position of the ZnGeN$_2$ layers.

8.2 Future work

8.2.1 Further development of ZnGeN$_2$ materials

8.2.1.1 Investigation of the transport properties of ZnGeN$_2$ on GaN

This book presented the preliminary results of unintentional carrier concentrations and mobilities for ZnGeN$_2$ films grown on sapphire substrates. Low carrier mobilities were attributed to the structural defects. The improvement in crystalline quality provided by closely lattice-matched GaN substrates is expected to result in improved carrier mobility in the ZnGeN$_2$ films. We propose an investigation of the effects of MOCVD growth parameters on carrier transport properties of ZnGeN$_2$ films grown on semi-insulating GaN templates as a part of future work. Temperature dependent Hall measurement complemented by the deep level optical spectroscopy (DLOS), deep level transient spectroscopy (DLTS) and depth resolved CL spectroscopy (DRCLS) will provide a comprehensive understanding of the defect energy levels and their physical origin in ZnGeN$_2$ films. The DLOS, DLTS and DRCLS measurements are ongoing work by collaborators.

8.2.1.2 Growth of high crystalline-quality ZnGeN$_2$ on sapphire by using buffer layers

Growth of ZnGeN$_2$ on GaN can be considered as homoepitaxy considering the similarity of crystal structure and close lattice-matching between these two materials. However, similar crystal structures and lattice parameters mean that the Bragg conditions of corresponding crystal planes of ZnGeN$_2$ and GaN would be similar, which makes it difficult to distinguish between the XRD scan profiles of the films from those of the substrate. The Bragg conditions for different polymorphs of ZnGeN$_2$ are already almost indistinguishable. For example, Pna2$_1$, Pmc2$_1$ and P6$_3$mc structures - all have XRD peaks at similar 2θ positions [36, 97]. Therefore, it will be more difficult to investigate the presence of cation disorder in ZnGeN$_2$ films if GaN is used as the substrates. Moreover, the very close band gap of ZnGeN$_2$ and GaN complicates the characterization of optical properties of ZnGeN$_2$ films grown on GaN.

For ZnGeN$_2$ films grown on sapphire, characterization of fundamental material properties is expected to be less complicated. However, the crystalline quality of ZnGeN$_2$ films grown on sapphire is affected by the large lattice mismatch between the film and the substrate [59]. A buffer layer (BL) can help reducing the effects of large lattice mismatch between the films and the substrate and thus, decreasing the extended defect densities, e.g., threading dislocation density in the film [132, 133]. The use of a BL has become an integrated part of GaN growth technique on sapphire. A similar approach may be useful for improving the quality of ZnGeN$_2$ films on sapphire substrates. Selection of a BL material (e.g., ZnGeN$_2$ or GaN) as well as optimization of BL growth conditions, growth rate, thickness and BL thermal annealing conditions would be required to achieve desired film quality [134].

8.2.1.3. Study the effects of cation ordering on optical properties of ZnGeN₂

The presence of cation disorder has been predicted to significantly lower the bandgap of $ZnGeN_2$ [25, 26]. The annealing study conducted on $ZnGeN_2$ on c-sapphire substrates, which has been presented in section 4.3 of this book, showed a blue shift in the absorption edge with increase in cation ordering. However, a thorough investigation of the effects of cation disorder on optical properties of $ZnGeN_2$ is still required. This study would require growth of $ZnGeN_2$ films with varying degree of cation disorder, which may be possible to achieve by varying the growth temperature [27]. One way to estimate the degree of cation disorder in the films is to compare the splitting of the super structure XRD 2θ-ω peaks. Raman spectroscopy can also be useful for this purpose.

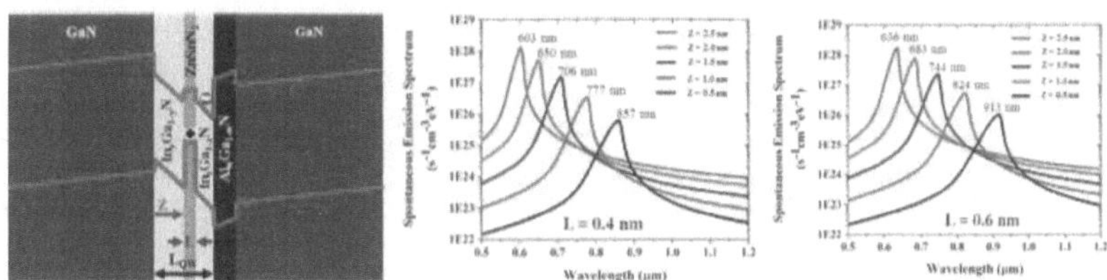

Figure 8.1 The effects of the position, Z, and thickness, L, of the $ZnSnN_2$ layer on the peak emission wavelength and spontaneous emission rate of a $GaN/In_yGa_{1-y}N/ZnSnN_2/In_zGa_{1-z}N/Al_wGa_{1-w}N/GaN$ QW structure.

8.2.2 Further development of InGaN/ZnGeN$_2$ heterostructure quantum wells

8.2.2.1 Reducing the cross-contamination in InGaN/ZnGeN$_2$ heterostructure QW

For manufacturing purposes, it is essential to control the impurities in InGaN layers and the composition of the ZnGeN$_2$ layers. Since unintentional incorporation is by nature indeterministic and uncontrolled, it is imperative to minimize the cross-contamination which will require a comprehensive understanding of the exact source of contamination. For example, different groups have identified the GaN template [122], reactor walls [121] as the source of unintentional Ga incorporation in MOCVD grown AlInN and the reactor geometry [120], reactor history [120], reactor pressure [120], metalorganic precursor flow rate [135] were found to influence the severity of Ga incorporation in AlInN.

The knowledge of the contributing sources and parameters will be useful in finding methods to minimize the cross-contamination. For example, a short interruption between the growth of the layers in the heterostructure might help reduce the level of cross-contamination.

8.2.2.2 Tuning the emission wavelength InGaN/ZnGeN$_2$ heterostructure QW LED by varying the position and thickness of the hole confining ZnGeN$_2$ layer

Numerical simulation suggests that the peak emission wavelength and peak spontaneous emission rate of InGaN/ZnGeN$_2$ and InGaN/ZnSnN$_2$ heterostructure based QWs are functions of the thickness and position of the ZnGeN$_2$ or ZnSnN$_2$ layer [35, 131]. As an example, the effects of the thickness and position of the ZnSnN$_2$ layer on the peak emission wavelength and peak spontaneous emission rate is shown in Fig. 8.1. Therefore, the position and thickness of the $ZnGeN_2$ layer can potentially provide an additional flexibility in tuning the color of the emission of the $InGaN/ZnGeN_2$ heterostructure QWs.

8.2.2.3 Additional characterization of QWs

The STEM images presented in in Chapter 7 of this book confirmed the position of thin $ZnGeN_2$ layer sandwiched between two InGaN layers. The CL and PL spectroscopy results showed effects of insertion of a think $ZngeN_2$ layer inside InGaN QWs. To better understand the emission properties of the grown $InGaN/ZnGeN_2$ heterostructure QW it is necessary to determine the composition of the cations in and the thickness of the $ZnGeN_2$ layer. Atom probe tomography measurements might be a useful tool for this purpose.

IQE is one of the commonly used metrics to evaluate the performance of QWs. For a certain excitation condition, a ratio between the PL intensities at low temperature and at room temperature can give an estimation of the IQE of the QW. The underlying assumption in this technique is that at low temperatures, the IQE becomes 100% due to the complete suppression of non-radiative recombination centers [136]. On the other hand, EQE of LED devices can be characterized by electroluminescence (EL) spectroscopy.

8.2.2.4 Fabrication of LEDs

In this book, PL and CL have been used to evaluate the emission properties of the $InGaN/ZnGeN_2$ heterostructure QWs. In PL and CL spectroscopy, the carriers are generated by a photon beam and an electron beam, respectively. However, LEDs are optoelectronic devices in which the carriers are generated by injected currents. The fabrication of final device which includes the growth of complete device stack including the p-type and n-type barriers and deposition of contact metals are part of the future work.